Architecture
木　建　筑
ISSUE 01　第 1 辑
SPRING　2017
刘　杰　主编

科学出版社
北　京

图书在版编目（CIP）数据

木建筑. 第1辑/刘杰 主编
—北京：科学出版社，2017.6
ISBN 978-7-03-053382-1

Ⅰ.①木… Ⅱ.①刘… Ⅲ.① 木结构-建筑设计-文集
Ⅳ.①TU366.2-53

中国版本图书馆CIP数据核字（2017）第134855号

责任编辑：吴书雷
责任印制：肖　兴／书籍设计：北京美光设计制版有限公司

科学出版社 出版
北京东黄城根北街16号
邮政编码：100717
http://www.sciencep.com

北京利丰雅高长城印刷有限公司 印刷
科学出版社发行　各地新华书店经销
*
版　次　2017年6月第一版
　　　　2017年6月第一次印刷
开　本　787×1092　1/8
印　张　40 1/2
字　数　750 千字
定　价　600.00元

（如有印装质量问题，我社负责调换）

Architecture
木 建 筑

▲ ISSUE 01　第 1 辑
▲ SPRING　2017

主办单位：　上海交通大学中欧木建筑研究中心
支持单位：　上海交通大学　欧洲木业协会
学术顾问：　（按姓氏笔画排序）
　　　　　　沈世钊　程泰宁

主　　编：　刘杰
编委会委员：（按姓氏笔画排序）
　　　　　　王兴田　阮昕（澳大利亚）　杨学兵　张绍明
　　　　　　吴培（瑞典）　赵川（日本）　赵辰
　　　　　　扬·索德林（瑞典）
编辑部主任：朱志军　吴书雷
执 行 主任：蒋音成　韩佳纹
中英文编辑：东鸿（美国）　高瑜　王恺

Host: Sino-European Wood Center
Supporter: Shanghai Jiao Tong University, European Wood
Consultant: Shen Shizhao, Cheng Taining

Chief editor: Liu Jie
Editorial committee: Wang Xingtian, Ruan Xing(Australia), Yang Xuebing, Zhang Shaoming, Pehr Mikael Sällström (Sweden), Zhao Chuan(Japan), Zhao Chen, Jan Soderlind (Sweden)
Editorial director: Zhu Zhijun, Wu Shulei
Executive director: Jiang Yincheng, Han Jiawen
Editor: Abraham Zamcheck(USA), Gao Yu, Wang Kai

主编的话
Editor's notes

历史地看，人类社会的发展的确呈螺旋形上升的态势，木结构建筑在欧洲、北美的发展也正印证了这一点。中国曾是世界上最大的、也是历史最悠久的传统木结构建筑使用国，直至今日，在东北林区、西南和东南山区，还有数量不少的井干式、穿斗式等传统木构建筑遗存。甚至在西南苗、侗、瑶、土家等少数民族聚居地，穿斗式的传统木构建筑营造方式还在延续着。

时间已经将人类文明带入了二十一世纪，传统木构建筑的余晖虽然还在这个世界上的局部地区艰难地维系着，时过境迁，它的热度与光芒早已随着传统营造的整体没落而逐渐衰微，其影响力和意义或已微存。与之相反的情形是，现代木构建筑在欧洲、北美等地区经过近百年的技术发展、不断进步后横空出世，凭借木材自然优美的纹理、温馨淡雅的色彩、清新怡人的香气、优雅端庄的造型以及循环共生的理念再一次征服世人！在全球一体化的低碳经济时代，强调人和自然和谐共生的生态文明社会建设进程中，现代木构建筑克服了传统木构诸多弊端之后，一扫木构建筑在二十世纪下半叶以来发展的颓势，犹如凤凰涅槃般浴火重生！这不是传统木构建筑简单的历史回归，更准确地说应是社会生态文明与现代工业文明结合的新生事物，体现的是当代工业先进制造技术与绿色生态的营造理念。

一定有人会问，是什么样的力量能让现代木构在发达国家崛起？

是世界应对气候变化的策略，是材料制造和建造技术的进步，是人类健康建筑体系的需求，也是崭新的营造理念使然！首先，全球气候变暖已经迫使低碳经济时代提前来临，全世界负责任的国家已经团结起来共同面对环境恶化带来的生态危机，这是现代木结构建筑发展的时代需求与文化大背景；然后，是现代木材工业高端的新型复合材料制造技术带来的与传统木材相比脱胎换骨式的革命性材料的出现，为现代木构建筑奠定了基础；第三，新时期人们对建筑体系的健康和舒适性要求的进一步提升；最后，是由于现代建筑结构技术的理论与设计实践在木构体系上的高度契合，也是二十世纪后半叶延续到二十一世纪的新时期创意设计的必然要求！

历史的轮回即将让木构建筑再度回到世界营造舞台的中心。追溯木构建筑体系的兴衰历程，我们能体察到人类营造文明的进化史。这一点对曾经普遍使用传统木构建筑的中国乃至东亚地区而言，或更具意义。千年前，宋代的中国已经出现了木构营造的标准化、模数化和预制装配式施工方式，并且影响了东部亚洲几百年。明代以后的中国，随着木材资源日益减少而逐渐没落。1949年以后，木结构这一先前最主要的建筑结构形式更是一度被主流营造活动所遗弃。改革开放以来，尤其是中国重新回到世界贸易大家庭后，得益于全球木材资源的支持，中国木构建筑市场逐步复苏。二十世纪八十年代以来，新颖别致的欧美现代木构建筑技术随着木材贸易西风东渐，登陆神州大地，与古老中国的木构文化再度重逢。最近，新一届政府明确提出了中华文化复兴的宏伟目标。木构建筑——这一承载着中国数千年建筑文明的载体，在当下强大工业文明的支撑下，也一定会再度回归，或可成为一支不可忽视的新兴力量。

近年来，中国政府日益重视现代木结构建筑在中国的发展，住建部和林业总局等相关部门组织行业专家和学者，相继完成了一系列木结构建筑工程或产品的国家标准的编制和再修编工作，初步建立起木结构建筑工程及相关产品标准体系，为现代木结构在中国的快速发展奠定良好的基础。2016年起，中国政府大力推广装配式建筑，发出"在具备条件的地方，倡导发展现代木结构建筑"（《中共中央国务院关于进一步加强城市规划建设管理工作的若干意见》2016年2月）的号召。旋即组织专家编制出台了《多高层木结构建筑技术标准》（GB/T 51226-2017，自2017年10月1日起实施）和《装配式木结构建筑技术标准》（GB/T 51233-2016，自2017年6月1日起实施）两部重要的技术标准，进一步完善了中国现代木结构建筑的技术标准体系，并解决了当前国内现代木结构建筑发展的技术瓶颈问题。至此，现代木结构建筑在中国的发展已经没有技术层面的障碍。国内也因此掀起了发展现代木结构建筑的热潮，如位于中国东北地区的吉林省，结合自身资源与产业状况，提出将木结构建筑产业列为未来全省重点发展的支柱产业。内蒙古自治区、贵州省等省区的部分地区也纷纷结合自身情况，制定加快发展现代木结构建筑产业的计划和措施。

在这样的一个时代，热爱木构建筑的人们，或将做些什么？

他山之石，可以攻玉。经过数年的酝酿和筹备，由上海交通大学木结构建筑研究团队与欧洲木业协会合作主办的《木建筑》杂志第1辑终于问世并呈现在您的面前。它的诞生可以弥补中国乃至第三世界国家当代建筑设计中建筑师对世界各国优秀木结构建筑的认识不足，也可以让喜欢木结构的业主或潜在业主们看到建造的更多可能，更可以让那些正在可持续发展理念下成长的建筑与土木专业大学生们拓展视野与思维。在这样一本模样普通的文本中，蕴含着来自世界各地现代木结构建筑作品的优秀案例，也展示着世界各地区、各国建筑师的不同设计文化与格调。期待无论作为何种身份的您，都会在这个文本中找到知音或钟爱的案例。

让所有热爱中国木结构建筑文化的朋友们，都参与到这个重要的历史进程中，一起用自己的智慧与努力来见证中华文化复兴的壮举！

Looking at history, the development of the human society is indeed a spiral up trend. The development of timber structure in Europe and North America also confirms this point. China was once the World's largest nation with the longest history of using traditional timber architecture. Until today, there are still a number of preserved log home style, column and tie style and other types of traditional timber structures in China's northeast forest region, southwest and southeast mountain regions. Even in the southwest regions with numerous minor ethnic groups, column and tie style traditional timber architecture construction is still practiced today.

Time has brought human civilization into the 21st century, and the afterglow of the traditional timber architecture still struggles to remain in some parts of this world. Over time, its glory continue to gradually decline along with the system of traditional construction methods, with little influence and significance. But contrary, modern timber architecture in Europe and North America has had near a hundred years of technological development, continued to progress since its birth. The natural and beautiful texture, warm and elegant colors and fresh and pleasant aroma of wood materials, this elegant and dignified style of wood construction continues to conquer the world. In the era of low carbon global economy with emphasis on sustainable and civilized development, modern timber architecture has overcome many drawbacks of traditional timber architecture, giving great opportunities for further development. This is not only just the rebirth of traditional timber architecture, but more specifically the combination of sustainable and industrial civilization. This reflects the advancement of modern industrial manufacturing technologies and eco-friendly, sustainable construction concepts.

One might ask: what were the reasons that drove the rise of modern timber architecture in developed countries?

It's the strategy to adjust to global warming; It's the advancement of material manufacturing and construction technologies; It's the demand for healthy building systems; It's the establishment of new concepts! Firstly, global warming has forced the beginning of the low-carbon economy era and all the responsible nations are united to face this eco-crisis caused by environmental deterioration. This is the era of modern timber architecture development with the necessary cultural background and conditions. Secondly, the advancement of material manufacturing techniques has provided the necessary foundation for modern timber structure with revolutionary new construction materials. Thirdly, this generation has higher demand for health and comfort in the building systems. Lastly, the theories and designs of modern building structure technologies are very fitting in modern timber structure systems, and it is also the needs of innovation in architectural designs for the 21st century that drove the development of modern timber architecture.

The cycle in history will bring timber architecture back to the center stage of the international construction industry. In retrospect of the rise and fall of timber architecture, we can see the evolution of human civilization. This has much significance to China and East Asia where traditional timber architecture existed. A thousand year ago in Song Dynasty, there were already standardized, modular and prefabricated construction methods that had impacted East Asian countries for hundreds of years. From the Ming Dynasty, construction of timber architecture slowly declined due to the decrease in forest resources. Since 1949, timber architecture, once the mainstream of construction industry, had been abandoned. However, since China's reform and opening-up policy was issued, especially since China joined WTO, the timber construction market had begun to recover, driven by the global timber trade. Since 1980s, novel timber structures from Europe and North America had emerged in China and met with the traditional timber structure culture. Lately, the new government has clearly proposed the goal of reviving Chinese culture. Timber architecture, the carrier of thousands of years of Chinese architectural civilization, will be revived with the strong support of industrialization, and become a force that cannot be ignored.

In recent years, the Chinese government strongly values the development of modern timber architecture in China. The Ministry of Housing and Urban-Rural Development and State Forestry Bureau gathered industry experts and academics to complete and revise a series of codes and standards for timber structure and products. The establishment of codes and standards provide the necessary foundation for the rapid development of modern timber architecture in China. Since 2016, the Chinese government strongly supported the development of prefabricated buildings and call the support to "encourage the development of timber structure in the suitable environments" (《中共中央国务院关于进一步加强城市规划建设管理工作的若干意见》2016 年 2 月). The government immediately organized expert groups to compile two important technical standards, the 《多高层木结构建筑技术标准》（GB/T 51226-2017, official implementation from 2017 October 1st）and the 《装配式木结构建筑技术标准》（GB/T 51233-2016, official implementation from 2017June 1st）; a step closer to better complete the technical standards system for China's modern timber architecture. At the same time, this resolved the current main technical challenges for the development of modern timber architecture in China. Until now, the development of modern timber architecture in China is finally cleared of obstacles in technical levels. As a result, there is a fast growing trend for the development of modern timber architecture in various regions. For example, the government of Jilin Province, located in northern China, had stated strong support for the timber construction industry in its region. With the combined local resources and industries, timber construction will become one of the most important industries in the province. In the Inner Mongolia Province, Guizhou Province and other areas also announced support for more rapid development of modern timber architecture industry tailored to each situational conditions in each area.

In this century, the loves of timber architecture, what will they do?

他山之石，可以攻玉。Through numerous years of gathering funding and preparation, the first edition of "mù Architecture", created in collaboration between the Timber Architecture Research Team from Shanghai Jiaotong University and the European Wood Association, is finally presented in front of you. The birth of this publishing can fill in the knowledge gap for timber architects in China and other developing nations with information of global and renowned timber architecture works. It also allows owners or developers, who have interest in timber architecture, to see the numerous possibilities of using timber structure. More so, it expands the perspectives and thinking of architects and civil engineering students who are growing in the generation of sustainable development. In this what appears to be a normal text, is stored the collection of excellent modern timber architecture works from around the world, while demonstrating the different design culture and style of these architects. I look forward that you, no matter your identity, may find a work from this publishing that you can admire and love.

For all the friends who have love and passion for Chinese timber architecture culture, lets participate in this important historical journey to witness the revival of Chinese culture with our knowledge and efforts.

上海交通大学中欧木建筑研究中心主任
建筑学系博士生导师
Director of Sino-European Wood Center
Doctoral Supervisor of architecture

2017 年 5 月

院士寄语
National academician's remarks

木结构建筑曾是中国古代建筑中的主角，经过数千年的发展，其应用范围之广，建造工艺之精，艺术成就之高，均达到无与伦比的程度，代表了中华传统文化的一个重要组成部分，且其国际影响遍及整个东亚地区。但自19世纪中叶以来，随着西方现代建筑技术及相应设计理念的传入，混凝土和钢材等新兴材料获得快速发展和应用，建筑行业经历了一个较大的变化过程。不仅传统的木结构建筑逐渐退出历史舞台，即使是基于"现代"技术的木结构建筑也没有获得快速发展，尤其是在一些大型的公共建筑领域。事实上，由于对森林资源的长期和过度开发，我国的木材资源逐渐匮乏，迄20世纪60年代后期，这一状况已相当严重，使木结构建筑基本上停止了发展。从那时起，全国大部分高校也相继停止开设木结构课程。相对于我国古代辉煌的木建筑文化，我们经历了一段悲惨暗淡的时期。

与此同一时期，欧洲、北美等地区的发达国家并没有停止木结构建筑的持续发展。丰富多样的材料品类和规格、富有时代性的设计理念、先进的建造技术，使各种新颖的现代木结构建筑获得广泛应用，展示了多彩的建筑形态，并充分体现了绿色生态建筑的一个重要发展方向。

随着改革开放的不断扩大和深入，这些先进的现代木结构建筑又逐渐传入我国，目前应用还不够多，业主、企业、设计人员等对它们的了解也还不够全面，说明它们在中国的发展仍处于初始阶段。但我国建筑业十分庞大，发展又非常快，而且目前正处于重要的转型升级阶段，通过不断创新向工业化、信息化、绿色化三位一体的现代化建筑业方向发展。这一宏观形势为现代木建筑的发展提供了巨大空间。（在这经济全球化的时代，原材料供应已不是问题，完全可以通过进口木材来解决我国的需求。）

在这一关键时刻，上海交通大学"中欧木结构建筑研究中心"联合国内外同行发起创办《木建筑》杂志，传播与交流国际上先进的木结构建筑信息，展示这类建筑体系在建筑美学、使用功能和绿色生态价值等方面的巨大优势，以期在中国大力推进这种先进建筑体系的发展，从而在新的技术与美学基础上复兴我们的木建筑文化传统，的确具有十分重要的意义。

这一杂志是国内乃至亚洲地区有关现代木建筑的第一本刊物，填补了这一领域的空白，的确办的正是时候，顺应了时代的需要。这体现了主编者的高瞻远瞩和巨大决心。我相信，这一杂志的成功创办必将为我国现代木建筑的健康快速发展起到重要的引领作用。我衷心祝愿《木建筑》杂志越办越好！

Timber structure was once the "main star" of ancient Chinese architecture. Over thousands of years of development, timber structure had reached an incomparable peak in terms of wide spread applications, delicate craftsmanship, and high levels of artistic achievement. It was a quintessential part of traditional Chinese culture, which also had strong influence on international regions such as the East Asia. With the introduction of modern western architectural techniques and related design theories since the 19th century, the applications of emerging building materials, such as concrete and steel, had spread rapidly and the construction industry transformed substantially. Not only was traditional timber structure disappearing gradually, but modern timber structure technologies also had not advanced, especially in the construction of large commercial and public buildings. In fact, the forest resource in China also became scarce due to long periods of over harvesting. In late 1960s, the depleting forest situation was so severe and the applications of timber construction essentially stopped due to limited resources. During this time, education on timber construction was removed in universities all over China. In comparison to the glorious development of timber structure of our Chinese heritage, this was a depressing period for the development of timber structures in China.

At the same time, advancements in timber structure continued in Europe and North America. With various types of products, dimensions and up-to-date design concepts, advanced construction techniques had been widely adopted in all types of modern timber structure. While demonstrating all types of architectural form, timber structure became an integral part of sustainable development and green buildings.

With the implementation of political reforms and open policies, modern timber structure had been introduced to China. However, the development of modern timber structure in China is still in the early stages since it has not been widely applied and recognized by clients, builders and designers. But, the large and fast-growing construction industry in China is entering a period of new advancements and upgrades with continuous innovations in industrialization, informatization, and modern green building technologies. This trend provides enormous opportunities for the development of modern timber structure. (In the age of globalization, raw material supply is no longer an obstacle. Global wood import is the solution to meet demands in China).

In this critical moment, the Sino-European Wood Center of Shanghai Jiao Tong University has taken the initiative to establish mù Architecture with domestic and overseas partners, aimed to spread the latest information on international timber structures and demonstrate the advantages of timber structure applications in terms of architectural aesthetics, functionalities, and green ecological values. This magazine holds significant value to promote the development of this kind of structural system in China and revive our traditional wood architecture heritage with advanced technologies and artistic foundations.

This magazine will be one of the first publishing on modern timber structures in China, filling a void in this industry. This is the right time to meet the demands of this generation. This publishing resembles the vision, ambition, and determination of the editor. I believe, this magazine will provide significant contributions to accelerate the development of modern timber structure in China.

2017年5月

沈世钊：哈尔滨工业大学教授、博士生导师，中国建筑学会木结构专业委员会顾问，中国工程院院士。
Shen Shizhao: Professor and Doctoral Supervisor at Harbin Institute of Technology, consultant of the timber structure committee of the Architectural Society of China, academician of Chinese Academy of Engineering (CAE).

院士寄语
National academician's remarks

　　不能设想离开"土木"后中国的传统建筑会是什么样子，甚至还会有传统建筑吗？无论是半坡遗址的木骨泥墙，还是河姆渡遗址的干栏建筑，或者是历经千余年仍旧屹立的南禅寺、佛光寺与应县木塔，它们都在诉说着"土木"的故事与情怀。《考工记》中的"殷人重屋，茅茨土阶"也是对土木建筑形象而生动的记载。

　　《康熙字典》中收录的木字旁汉字有 1413 个，其中与建筑有关的多达 400 多个，诸如梁、柱、椽、檩、栋、楼等字；还有很多与生活密不可分的木字旁文字，诸如桌、椅、床、枕等，传统建筑中大木作的结构与小木作的装饰是基本法则，而且也是传统建筑的精华所在。植树为林，伐木为材，立柱架椽，广宇重栋，造屋筑园，这些老屋（可能并无精巧的艺术造型）不再仅仅是居住的房子，而是祖辈生命和血脉的传承，更是天人合一精神的践行与升华。

　　我国近代由于资源和技术的原因，也由于受到西方建筑与文化的影响，在近代的我国，木结构房屋已繁华不在，甚至被弃之如敝屣。过去几十年，由于林业资源的匮乏和木材的短缺，政府对木材在建筑上的应用制定了严格的限制措施，提倡以钢代木、以塑代木，也使木结构房屋一直被排除在主流建筑之外。这与西方国家的情况形成了明显的对比，在木结构的研究和应用方面，近年来不仅应用原生态建筑较为广泛的北欧国家发展很快，而且美国、加拿大、日本等国也有长足发展，这些国家每年新建的住宅中有超过半数的房屋采用了木结构，并且有继续上升的趋势。同时，越来越多由木材构筑成的单一建筑、集合住宅和大型公共建设项目纷纷跃升到国际舞台，木建筑俨然成为二十一世纪的建筑新宠。

　　最新研究显示，木材不仅耐震抗压，利用新技术提高防火性能后，使得木材优点越发突出，而且木材属天然的再生材料，因此具有节能、环保、益于人体健康等优点，完全符合"绿色建筑""生态建筑"的可持续发展的要求。因此，吸取国外的经验，加强对木材研究和开发，是一个值得我们特别重视的课题。

　　出于一种前瞻性，刘杰博士很早就关注木建筑的研究。五年前，在上海交大成立了"木建筑研究中心"，最近又倡导出版刊物——《木建筑》。这是一项极富开拓性的工作。特别是《木建筑》的出版，填补了我国在木建筑研究方向的空缺，它将能帮助我们深入吸取国外经验，进一步推动我国木建筑的研究、开发和应用。我衷心期望，通过刘杰博士的推动和《木建筑》的出版，使更多人关注"木建筑"在我国的发展，中国传统文化中的重要部分——木建筑，能够提升创新，重放光彩。

It is hard to imagine what traditional Chinese architecture will be without mud and wood, or if any traditional architecture will exist at all? Whether it is the wooden-stud mud walls of the Banpo Site, the stilt houses in the Hemudu Site, or the wooden structures that remained standing over a thousand years such as the Nanchang Temple, Foguang Temple and the Pagoda, they all tell a story about mud and wood. From the section on building of houses in the "Kao Gong Ji" (The Records of Examination of Craftsman), a classic work on science and technology in Ancient China written sometime during the late Spring and Autumn period (771 to 476 B.C.), is a powerful record to demonstrate the importance of wood and mud architecture.

The "Kangxi Dictionary" has 1413 Chinese characters that contain the woodcharacter(木)as a component. Among these characters, there are over 400 characters associated with buildings such as beam (梁), column (柱), rafter (椽), purlin (檩), center purlin (栋), and floor (楼). There are also many characters associated with furniture and daily necessities products with the wood character (木)such as table (桌), chair (椅), bed (床), and pillow (枕). Whether it's large structural components or small decorative and furniture, it is fundamental to include the wood character (木) in all characters associated with traditional architecture. From harvesting to construction, the applications of wood were essential part of historical architecture. These old buildings are not only just shelter or art pieces, but also represent the blood, sweat, and tears of our ancestors and the essence of our cultural heritage.

Because of resources, technologies, and influences from western architecture and culture, the applications of timber structure gradually disappeared in our nation during recent periods. Over the past few decades, the government restricted applications of wood materials for building construction due to the nation's depleting forest resources. As a result, steel and plastics replaced wood entirely and timber structure was neglected from standards in the building industry. This situation was completely opposite to nations such as countries in Europe, the United States of America, Canada, Japan and etc, where construction of sustainable buildings became more and more popular. The research and applications of timber structures had continuous development with over half of all residential developments every year were built with wood in these countries. At the same time, more and more public buildings and commercial developments on the international stage also used wood as primary structural material. Timber structure has become the new favorite of the 21st century building industry.

Recent studies indicate that wood has high seismic and compressive strength performances. With new techniques to enhance the fire resistance of wood, applications with wood gained significant breakthrough. Not only is wood a natural and renewable resource, it also provides energy saving, eco-friendly, and health benefits when used as a building material. It completely meets the requirements of sustainable development and green buildings. Therefore, we must learn from foreign experiences and enhance the research and development of wood applications. This is a subject we must pay close attention to.

With a forward-looking perspective, Dr. Liu has put focus on the research of timber structure very early. Five years ago, the Sino-European Wood Center was established at Shanghai Jiao Tong University. And recently, the mù Architecture magazine will soon be published. This is a pioneering work to fill in the blanks of timber structure studies in our country. This magazine will help us learn more from overseas experience and push forward our country's research, development, and application of timber structure. Through Dr. Liu's efforts and the publication of mù Architecture, I sincerely hope that more people will pay attention to the development of timber structures in our country. As an essential part of traditional Chinese culture, timber structure can drive innovation and brilliance.

2017 年 5 月

程泰宁：东南大学建筑设计与理论研究中心主任、教授、博士生导师，筑境设计主持人，全国工程勘察设计大师，中国工程院院士。
Cheng Taining: Chairman, Professor, and Doctoral Supervisor of the Architectural Design and Theory Research Center, Southeast University, Leader of CCTN, Master of Engineering Survey and Design (awarded by MOC), academician of Chinese Academy of Engineering (CAE).

目录
Contents

主编的话	i	Editor's notes
院士寄语	iv	National academician's remarks
院士寄语	vi	National academician's remarks

简讯 Brief

爱德华·金主教教堂	001	Bishop Edward King Chapel
大胆的法式设计	006	Daring French design
多功能	014	Multifunctional
再生木的直角设计	021	Right-angled design in recycled wood

在循环木材的怀抱中入眠	029	Sleep well in the embrace of recycled wood
阿塞拜疆的宁静岛屿	037	Tranquil islands in Azerbaijan
向数学致敬	044	Tribute to mathematics
非比寻常的点心铺	051	Unusual cake shop by Kuma

办公教育建筑 Office & Education Building

个人与工业	056	Individual & industrial
环抱木材的幼儿园	070	Preschool embrace wood
平衡于隐私与社区的设计	089	Privacy and neighbours informed design
简单的材料体现想象力和对称	106	Imagination & symmetry come together in material simplicity

宗教建筑
Religious Building

松木暖热未来主义新地标	121	Pine warms futuristic landmark
建筑与自然之间日式和谐	134	Japanese harmony between building and nature

文化建筑
Cultural Building

建筑师是书的推手	144	Architects promote books
世博2015——米兰之木	153	Expo 2015 – Wood in Milan
钻石声效	168	Diamond acoustics
李伯斯金在柏林	180	Libeskind in Berlin

高层木结构
High Rise

人民之家高层公寓	192	Folkhem's block
城市木业进展	205	Urban development in wood

专题
Specialty

2016年瑞典木建筑奖提名项目	220	The nominated for the 2016 Swedish Timber Prize
升华建筑物的第五立面	230	Raising the profile of the building's fifth facade

对话 Voice

访谈1：隈研吾	244	Interview 1: Kengo Kuma
访谈2：王兴田	246	Interview 2: Wang Xingtian

中国实践 Practice in China

胶合木在当代佛教建筑中的设计应用
——杭州香积寺复建规划设计 　　251　　The Design Application of Glued Laminated Timber in Contemporary Buddhist Architecture
Hangzhou Xiangji Temple Rehabilitation Planning Design

"土·木"再释
——深圳隐秀山居建筑设计中的"土·木"运用与实践 　　266　　Re-Considering "Earth" & "Wood"
Application and Practice of "Earth & Wood" in the Architectural Design of the Castle Hotel-Genzon in Shenzhen

成都毗河项目经验分享 　　285　　Chengdu Pi River Project Experience Sharing

云建筑
——万科青岛小镇游客中心诞生记 　　298　　Cloud Building
Vanke TsingTao Pearl Hill Visitor Center

Bishop Edward King Chapel

爱德华·金主教教堂

校译：蒋音成
摄影：Niall Mclauglin

| Bishop Edward King Chapel in Cuddesdon, Great Britain by Niall McLaughlin
| Published 28 November 2013
| 爱德华·金主教教堂，英国卡兹登，尼亚尔·麦克洛宁
| 2013年11月28日出版

HOW DO YOU make a brand new building spiritual?

When London-based Niall McLaughlin Architects were commissioned to design Bishop Edward King Chapel, near Oxford in England, they chose to give the building an elliptical form and blend soft and hard materials in light colours.

The interior boasts a treelike timber construction of prefabricated glulam that stretches from floor to ceiling. Its branches create a second ceiling in the form of cathedral vaulting. The posts rising up also define the space in what is otherwise a quite open-plan design. Larch and ash have been used for all the interior fittings, dominating the look of the chapel.

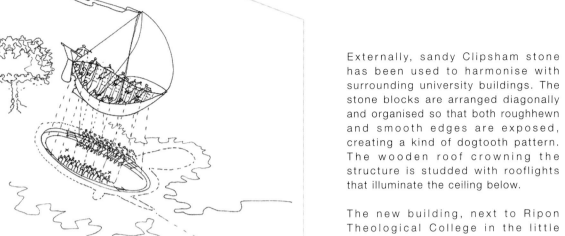

Externally, sandy Clipsham stone has been used to harmonise with surrounding university buildings. The stone blocks are arranged diagonally and organised so that both roughhewn and smooth edges are exposed, creating a kind of dogtooth pattern. The wooden roof crowning the structure is studded with rooflights that illuminate the ceiling below.

The new building, next to Ripon Theological College in the little village of Cuddesdon in the county of Oxfordshire, will primarily be used by students at the college and nuns from the local religious order. The chapel took 18 months to build.

Bishop Edward King Chapel
爱德华·金主教教堂

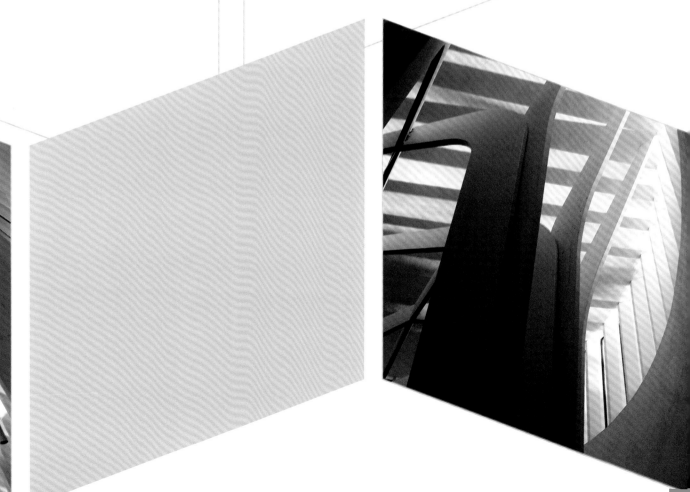

爱德华·金主教教堂
Bishop Edward King Chapel

您将如何让新建筑赋予灵魂？

当总部位于伦敦的尼亚尔·麦克洛宁建筑师事务所接到在英国牛津附近建造爱德华·金主教教堂的任务时，他们选择混合软硬材质设计一幢椭圆形的浅色建筑。

室内大胆采用由预制胶合木搭建的树形木结构，从地板一直延伸至顶，其分枝又搭建了一个教堂拱顶形状的天花板。同时，向上延伸的树形柱重新诠释了完全开放式的空间设计。落叶松木和白蜡木也运用于所有室内配件，是教堂的主要视觉元素。

室外为了与周边学校建筑相协调，采用克里普歇姆砂石，石块整齐斜错地对角垒砌，既粗犷又平滑的边缘毫无掩饰地显露出犬齿般的效果。冠于建筑物之上的木屋顶上镶有照明灯，从天花点亮整个空间。

这个新建筑在里彭神学院旁，位于牛津郡的一个名为卡兹登的小乡村，将主要被学院里的学生和当地修道院的修女使用。此教堂建设周期为18个月。

Daring French design
大胆的法式设计

校译：蒋音成
摄影：Ternisien

| La Maison-vague in Reims,
  France by Patrick Nadeau
| Published 11 September 2014
| 法国兰斯的波浪住宅，帕特里克·纳多设计
| 发表于 2014 年 9 月 11 日

A LAYER OF grass, herbs and flowers adorns the roof of this rounded house by architect Patrick Nadeau. He calls his work "La Maison-vague", which translates as "The wave-shaped house". The building is located near the city of Reims in north-east France and is the first in a project totalling 63 experimental homes in the area.

Daring French design
大胆的法式设计

建筑师帕特里克·纳多设计的这座弧顶住宅，屋顶披着一层鲜花绿草，熠熠生辉。他称其为"La Maison-vague"（法语），意思是"波浪形的住宅"。该建筑靠近法国东北部的兰斯市，这里有一个由63栋试验性住宅构成的项目，这座建筑正是其中的第一栋。

Daring French design
大胆的法式设计

A structural frame in cross-laminated timber has been used to create the wavy profile. The building stands entirely on a raised wooden deck that lifts it above the ground. Only the foundations are concrete. The location, vegetation and double-walled facades ensure its thermal performance. The outer walls are made from transparent polycarbonate, and the inner walls from untreated pine cladding. The green sedum roof provides protection from seasonal heat and cold.

波浪形的外形有正交胶合木做成的结构框架。建筑底部由木甲板平台垫起，建筑物立于其上，只有基础为混凝土。选址、植被和双层外墙保证了建筑的保温性能。墙体外侧使用透明的聚碳酸酯材料，内侧则采用未做防腐处理的赤松饰面。绿色景天植物屋面能很好地调节由四季冷热变化给建筑带来的影响。

Daring French design
大胆的法式设计

多功能
Multifunctional

校译： 蒋音成
摄影： Inazumi

| 日本神户和谐体育馆，竹中公司设计
| 发表于 2013 年 12 月 20 日
| Harmonie Hall in Kobe, Japan by Takenaka Corp
| Published 20 December 2013

ENORMOUS ROOFLIGHTS FACING north and south bathe this sports hall in light. The building is located in Kobe, Japan, and is used as a basketball court and auditorium for the pupils of Kobe International High School.

1. Junior high school
2. Gymnasium
3. High school
4. Boarding house
5. Seminar room
6. Harmonie Hall

该建筑坐落于日本神户，是神户国际高中的一座篮球场和礼堂。南北两侧的巨大高窗，让这座建筑沐浴在自然光中。

多功能
Multifunctional

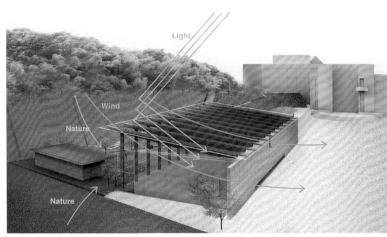

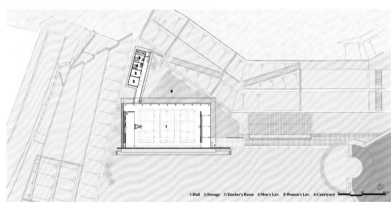

1:Hall 2:Storage 3:Teacher's Room 4:Men's Lav. 5:Women's Lav. 6:Courtyard

体育馆建筑通常表现为封闭的空间，与周边环境相隔离。但这个项目却截然不同。通过展现建筑的实木结构框架，建筑物得以与环绕校园的森林和谐统一。

Sports halls tend to be enclosed spaces divorced from their surroundings, but that is not the case here. By exposing the structural frame in solid wood, the building harmonises beautifully with the forest that surrounds the school campus.

The sports hall was made in wood and concrete to blend in with the school's other buildings. A 46 metre long concrete wall was erected to the south to form the hall's entrance. Narrow windows line the top and bottom of the wall, allowing floods of light into the interior.

From the north, the sports hall almost looks like part of the landscape, thanks to the windows that make up a large part of the facade.

多功能
Multifunctional

这座体育馆采用木材和混凝土建造，以融入学校的其他建筑之中。南面一面46米长的混凝土墙形成了主入口，墙上窄条形的高窗和地窗将自然光引入室内。

从北侧看这座体育馆，建筑仿佛就是景观的一部分，这得益于这些窗户，它们构成了立面的主要部分。

校译:蒋音成
摄影:Shigeo Ogawa

| 木结构 1 和 2,日本东京,FT Architects
| 2014 年 3 月 6 日发行
| Timber structure I & II in Tokyo, Japan by FT Architects
| Published 6 March 2014

再生木的直角设计
Right-angled design in recycled wood

位于东京日本工学院校园内的2座新建体育馆是由日本FT建筑师事务所操刀设计的,致力于打造能够激发学生活力的空间。全部是用本地木材设计的低成本无柱空间。其解决方案是,用其他建设项目的废旧木材,搭建2个自承力顶的大厅。

JAPANESE FIRM FT Architects is behind the design of two new sports halls at the Kogakuin University campus in Tokyo. The brief was to create accessible and inspiring spaces in which the students could be active. They were to be free from pillars and columns, cheap to construct and made from local timber. The solution was to create two halls with self-supporting roof structures in timber recycled from other building projects.

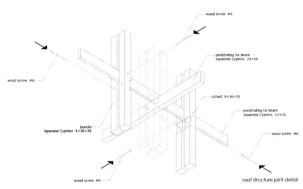

再生木的直角设计

To achieve pillarless spaces using cheap construction methods, it was necessary to come up with innovative new ideas on how wood can be used. The key to success was to create a lattice of wooden frames that employ slender offcuts from a local furniture maker. Ensuring that each lattice was entirely made up of right-angled components required minute accuracy during assembly. Nuts and bolts hold the wooden frames together.

The other hall is used for boxing training and is built along the same lines. However, it comes across as less light in tone. Its lattices are stepped in line with the pitched roof and the wood is much chunkier.

再生木的直角设计

　　为了实现低成本无柱空间的效果，必须创造出木材运用的新思路。从当地家居生产商那里淘来的细长边料，搭成木结构格珊，这是成功的关键。必须确保组装时，每一个框格都是由精准直角的部件搭建而成。木框架用螺栓和螺母做固定。

　　同排建造的另一个馆用于拳击训练，但色调稍暗。其格珊沿屋顶而上承人字形，木材更为粗犷。

图片所示的空间是弓道馆,一种日本的射击武术,旁边空间是同样7.2米乘以10.8米尺寸的另一个馆,采用同种建构原理。但是两个馆因为不一样的结构和空间设计而给人截然不同的感受。

The hall pictured above is used for Kyudo, the Japanese martial art of archery. Next to this sports hall is another hall of identical size, 7.2 metres x 10.8 metres, and built according to the same construction principles. And yet the halls feel completely different due to their different structural and spatial designs.

Right-angled design in recycled wood

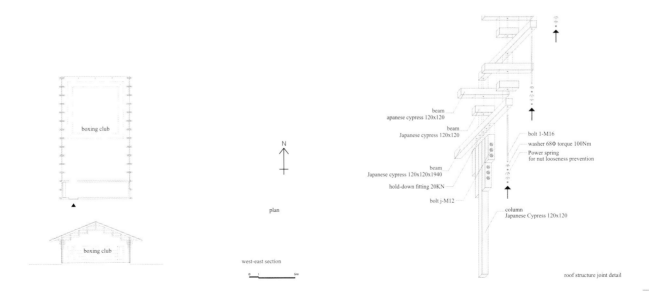

boxing club

boxing club

plan

west-east section

beam
apanese cypress 120x120
beam
Japanese cypress 120x120

beam
Japanese cypress 120x120x1940
hold-down fitting 20KN
bolt j-M12

bolt 1-M16
washer 68Φ torque 100Nm
Power spring
for nut looseness prevention

column
Japanese Cypress 120x120

roof structure joint detail

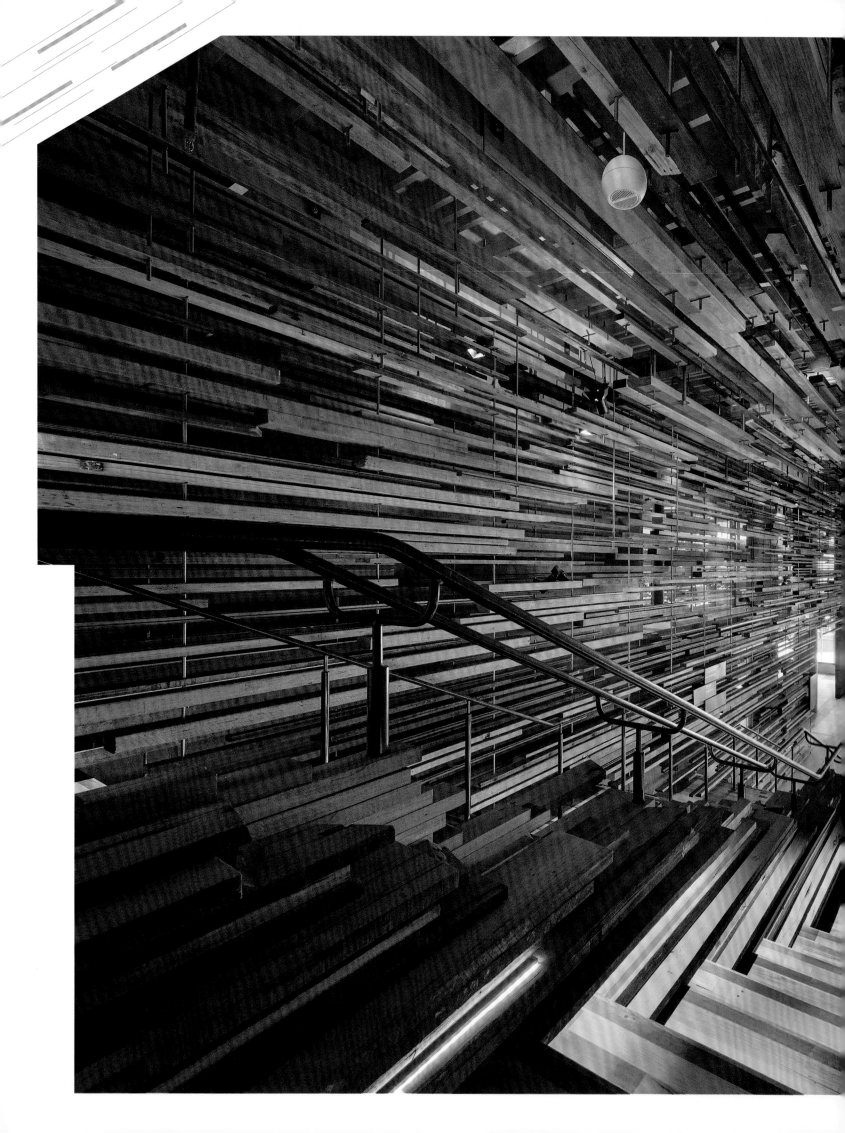

Sleep well in the embrace of recycled wood

在*循环木材*的怀抱中入眠

校译：蒋音成
摄影：John Golings

| 澳大利亚堪培拉 Hotel 酒店，三月工作室设计
| 发表于 2014 年 9 月 12 日
| Hotel Hotel in Canberra, Australia by March Studio
| Published 12 September 2014

Sleep well in the embrace of recycled wood

WALKING THROUGH THE lobby of Australia's Hotel Hotel is like swimming in a sea of wood. The ceiling structure alone comprises 2,150 strips of wood. Thousands more pieces of recycled wood line the walls, creating irregular patterns around the building's prefabricated concrete pillars.

走进澳大利亚 Hotel 酒店大堂,仿佛畅游在木材的海洋里。仅屋顶结构就由 2150 根木条构成。成千上万根循环利用的木板条排成墙,营造出不规则的图案将预制混凝土柱包裹起来。

A total of 1,200 steel rods hold the material in place. With the timber coming in different widths and sizes, no two surfaces are the same, each being individually formed and placed at varying distances from each other. More than 50 artists, designers and craftspeople were brought in to create the eclectic interiors for this hotel, of which the lobby is just one example.

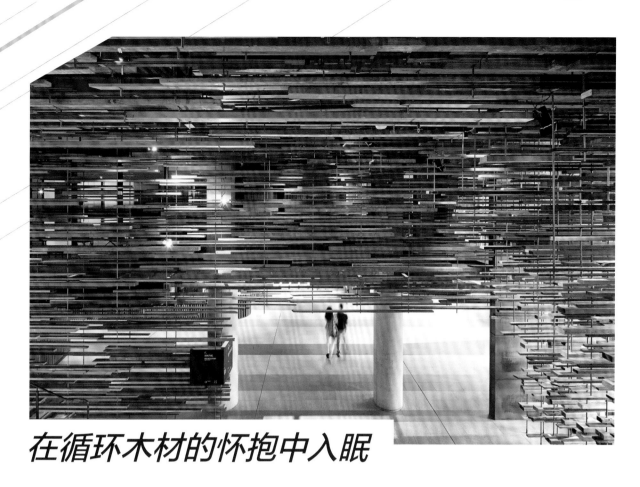

在循环木材的怀抱中入眠

这些木条由 1200 根钢杆固定得恰到好处。由于木条宽度和尺寸不一，任何两个表面都不雷同，木条之间以不同的间距排列得错落有致。有超过 50 名艺术家、设计师和工匠共同来为酒店创作兼容并蓄的室内空间，这个大堂只是其室内设计的一例。

March Studio designed the lobby and the magnificent staircase, which was formed by arranging stacks of recycled planks next to each other. All the wood has been gathered from nearby building sites and much of the timber comes from the blackbutt tree, which grows in the coastal forests of New South Wales and Queensland.

三月工作室运用这种木板条错落层叠的方式，设计了酒店的大堂和一个壮观的楼梯间。所有的木材都来自附近的建筑场地，其中大部分是长在新南威尔士和昆士兰海岸森林的黑基木。

阿塞拜疆的宁静岛屿
Tranquil islands in Azerbaijan

校译：蒋音成
摄影：Kerem Sanliman, Sergio Ghetti

| 阿塞拜疆巴库盖达尔·阿利耶夫国际机场，由 Autoban 设计
| 发表于 2014 年 11 月 25 日
| Heydar Aliyev Int. Airport in Baku, Azerbaijan by Autoban
| Published 25 November 2014

EACH YEAR, 6 million travellers pass through Heydar Aliyev International Airport in the Azerbaijani capital of Baku. Now, as they enter the airport, travellers are met by 16 cocoons made from interwoven panels of oak plywood.

每年有 6 百万旅客从阿塞拜疆首都巴库的盖达尔·阿利耶夫国际机场过境。如今，当人们走进机场，迎接他们的是 16 个由橡木胶合板"编织"出的"虫茧"。

Tranquil islands in Azerbaijan

The cocoons create tranquil little islands in an otherwise hectic environment. The garlic bulb-shaped rooms house cafés, a champagne bar, a beauty salon, a bookshop, a play area and a left-luggage area.

在机场纷繁嘈杂的环境里,这些"虫茧"形成了一个个宁静的岛屿。蒜头形状的空间里有咖啡厅、香槟吧、美容室、书店、游戏区和行李寄存处。

阿塞拜疆的宁静岛屿

阿塞拜疆的宁静岛屿

The structures were made in Ankara and erected on site. The architectural practice behind the design is Autoban, which has its headquarters in Istanbul. Architect Seyhan Ozdemir explains that they wanted to break the conventional norms for the look of an airport. The goal was to create a warm, calm and welcoming environment. According to Ozdemir, wood, stone and textiles were key materials in achieving this feel. The travelling public's response to the garlic bulb cocoons has been overwhelmingly positive.

这些结构在安卡拉制造，现场拼装。建筑设计是由总部位于伊斯坦布尔的 Autoban 公司完成的。建筑师 Seyhan Ozdemir 阐述了他们打破机场固有面貌的想法，目的就是营造出温暖、安静、热情欢迎旅客的环境。他说，木材、石材和织物是达到这种效果的关键材料。旅客们对这些蒜头形"虫茧"也是好评如潮。

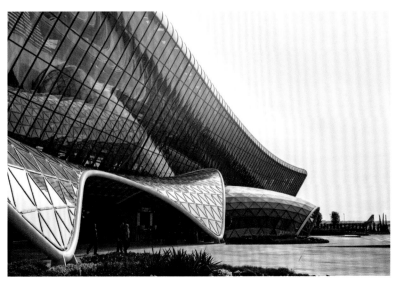

数学向致敬
Tribute to Mathematics

校译：蒋音成
摄影：Jens Lindhe

| 加梅尔·赫勒鲁普高中的赫勒鲁普体育馆，丹麦 BIG 建筑事务所
| 2014 年 3 月 12 日发行
| Hellerup in Gammel Hellerup, Denmark by BIG
| Published 12 March 2014

WHEN THE HIGH school in Gammel Hellerup, Denmark, needed a new sports hall, they also ended up with a brand new schoolyard. The hall by the architects BIG was opened in 2013. It has been sunk down at least five metres into the ground between the school's existing buildings. Its vaulted wooden roof has been created using a number of curved glulam beams that span the 1100 square metre hall. At the same time, the roof creates a brand new outside space in the schoolyard above. Untreated oak has replaced the previously asphalted outdoor surface. The wooden structure rises out of the ground like the back of a whale, offering plenty of space to sit and chat.

The idea of creating a wavy roof arose out of a handball match. Architect Bjarke Ingels at BIG drew inspiration from the arc of the handball through the air and sought to replicate that movement in the roof joists.

"In homage to my old maths teacher, we used the mathematical formula for a ballistic curve to create the geometry of the roof," he explains.

Tribute to Mathematics

CENTER OF ATTENTION
But how do we create a new multi-purpose hall in the courtyard without blocking the view of the existing buildings?

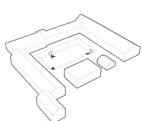

CONNECTIONS
The hall is connected to the existing buildings underground.

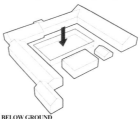

BELOW GROUND
The new multi-purpose hall is placed 5 meter underground to keep the courtyard intact and minimize shading on the surrounding buildings.

ROOF
The roof is curved to maximize the space for different sports activities.

SURPLUS SOIL
The surplus soil be moved to an outdoor sportsarea.

ROOF ABOVE GROUND
The roof above ground as seen in the courtyard.

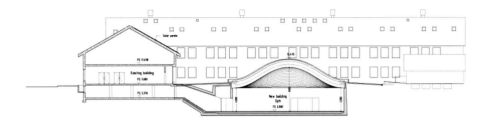

丹麦的加梅尔·赫勒鲁普高中急需一个新的体育馆，但校园里已经没有多余的地方可以用了。这座由 BIG 建筑事务所操刀的体院馆于 2013 年投入使用，面积 1100 平方米，位于现有教学楼间的庭院下沉 5 米余深的地方，巧妙地运用数根曲形胶合木梁支撑起拱形木质屋顶，打造了一个全新的校园地面空间。原始橡木甲板取代原先的沥青地面。高出地面的木结构宛如鲸鱼背脊，提供开阔的休闲空间。

这个波浪屋顶的灵感来源于一次手球比赛。BIG 事务所的设计师比雅克·英格斯尝试将手球在空中划出的弧形轨迹凝固于屋顶龙骨之上。

"为向我以前的数学老师致敬，我们用数学公式测算出的弹道弧线来打造几何屋顶。"他补充道。

Mathematics 数学

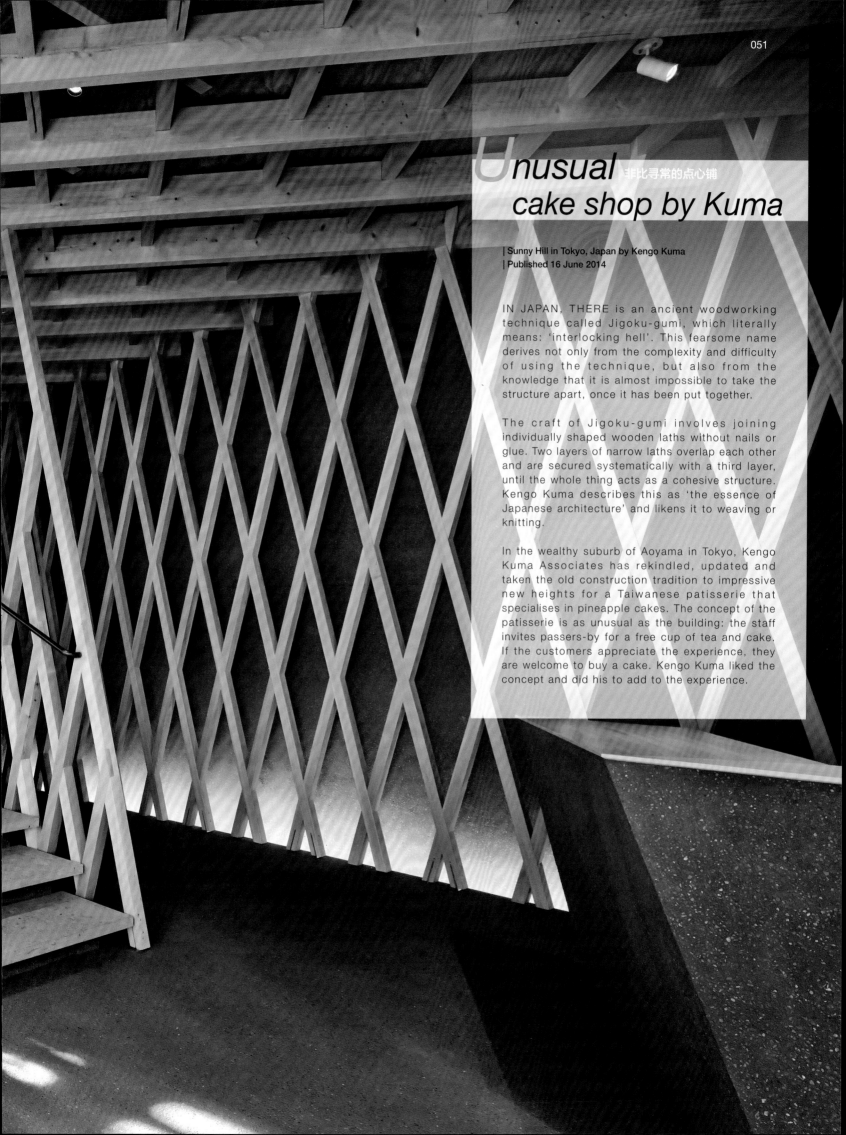

Unusual cake shop by Kuma

非比寻常的点心铺

| Sunny Hill in Tokyo, Japan by Kengo Kuma
| Published 16 June 2014

IN JAPAN, THERE is an ancient woodworking technique called Jigoku-gumi, which literally means: 'interlocking hell'. This fearsome name derives not only from the complexity and difficulty of using the technique, but also from the knowledge that it is almost impossible to take the structure apart, once it has been put together.

The craft of Jigoku-gumi involves joining individually shaped wooden laths without nails or glue. Two layers of narrow laths overlap each other and are secured systematically with a third layer, until the whole thing acts as a cohesive structure. Kengo Kuma describes this as 'the essence of Japanese architecture' and likens it to weaving or knitting.

In the wealthy suburb of Aoyama in Tokyo, Kengo Kuma Associates has rekindled, updated and taken the old construction tradition to impressive new heights for a Taiwanese patisserie that specialises in pineapple cakes. The concept of the patisserie is as unusual as the building: the staff invites passers-by for a free cup of tea and cake. If the customers appreciate the experience, they are welcome to buy a cake. Kengo Kuma liked the concept and did his to add to the experience.

隈研吾：
非比寻常的点心铺

Unusual cake shop

校译：蒋音成
摄影：Daici Ano

| 东京"微热山丘"凤梨酥铺，日本隈研吾
| 2014年6月16日发行

在日本，有一种木工古技艺叫做"地狱集"，字面意思为：锁扣地狱。这个恐怖的命名不仅来源于使用此技术的复杂性和其难度，并且一旦组装开始，就很难拆解。

"地狱集"技艺包括不用钉子和胶水连接一个个成型的木板条。将第一层和第二层窄板条相互重叠，再用第三层板条系统地进行固定，直到整体形成紧密相连的建筑体。隈研吾认为此技艺是"日本建筑的本质"，可以比拟为编织。

在东京富裕的青山郊区，隈研吾建筑都市设计事务所为一家台湾凤梨酥铺重拾并升级这项传统技艺，创造了新高度。这家凤梨酥铺的理念也同这座建筑一样非比寻常：店员邀请路人免费品尝乌龙茶和凤梨酥。如果顾客享受此体验，便会满意地购买凤梨酥。隈研吾欣赏这种理念，也将他的想法融入此体验。

东京"微热山丘"凤梨酥铺，
日本隈研吾

Individual & industrial
个人与工业

Text: Camilla Schlyter | 校译：韩佳纹　朱志军
摄影：Norman Radon

| Illwerke Centrum Montafon in Vorarlberg, Austria by Hermann Kaufmann
| Published 17 September 2014

到四。但是如果我们算上减少的能耗及更短的工程时间来看，木建筑从长远的角度上来看是更便宜的。"因此，客户们需要对于木结构的优势有更清晰的理解。比如说，迪特里希（Dietrich）认为，用来制造建筑材料的能源需要被考虑进来。在瑞典，他们正积极努力鼓励客户和建造业，让他们从一个全生命周期的角度来看待这些最常见的材料。这样做的目的在于，在不久的将来，他们可以将碳排放量、生产能耗、运输排放和原材料的可回收性等因素纳入考量。

考夫曼（Kaufmann）认为客户、开发商、建筑师、施工人员以及工业之间的影响深远的关系，可以创造一个共享的知识平台，一个实践和理论并存的木建筑文化平台。这为让所有人的意见都被聆听与尊重打下了在对话中寻求解决方式的基础。Kaufmann 和迪特里希（Dietrich）认为，至少要让木建筑更加划算，所以罗列出对木建筑施工的各项标准是十分重要的，这些在建筑水泥产业里已经逐渐完善。那么，该如何设计一个可以同时用于多个项目，又能保证质量的要求，还能创造独特的建筑物的生产体系呢？"奥地利大多数的木建筑是以项目细节和设计建造的，但是各个不见都是工业化制造的。" 赫尔曼·考夫曼（Herman Kaufmann）说。

怎样在建造过程之中，既可以降低成本，又能创造高质量的独特的木建筑？建筑师卡米拉·斯里特（Camilla Schlyter）带你前往奥地利福拉尔贝格州的发现。

对木构建筑的将来有着最重要的影响因素是什么？两名奥地利建筑师赫尔穆特·迪特里希（Helmut Dietrich）和赫尔曼·考夫曼（Herman Kaufmann）毫不犹豫地回答："通过高水平的预制来创造高质量的建筑物。"如果想要追寻品质和经济性同样重要的，木材的工业化建造，对于链条当中的所有环节，都需要对品质有相同的野心和追求。"我们尝试对于所有的东西达成一致意见，从木材的评分、与林业的关系、什么样的原木可以用在建筑的哪一部分上，同时也包括我们如何运用对于建造木住宅的传统知识。这需要结合现代高科技，针对一些被建筑师和手工艺人拥护的传统木技艺的新研究成果，及工匠和施工公司已经建立的工业生产方法。"Herman Kaufmann 说。赫尔穆特·迪特里希（Helmut Dietrich）继续说道："在奥地利，用木头建造房屋比用混凝土贵百分之三

如果找寻一个完美的例子，可以去赫尔曼·考夫曼（Herman Kaufmann）的新作品，位于蒙塔丰，福拉尔贝格州的 Illwerke Centrum，它实现了高质量工业建造，也是世界上最大的混合木结构办公楼之一。这座建筑有 120 米悬挑伸出于一个人工湖之上，5 层的结构在 6 个半星期就搭建完工。这是一个在形式和尺度上，简单又特殊的建筑物。它更让人联想到一件精美的家具，而不是一栋能源公司的现代写字楼。它对细节的追求和质量都令人赞叹。立面使用的是未加工过的原木，因为考夫曼（Kaufmann）不愿使用任何形式的防火处理。他认为，未经加工的木材可以在它的生命周期的最后被回收，而且能够优雅地老去。建筑立面的横向突起有着很多实用的考量，比如这样便于窗户的清洁、防晒及提高防火性。但是，这样的设计根本地还是一个建筑设计上的考量。考夫曼（Kaufmann）没有采用他惯用的当地云杉，而是选择了更加合适的橡木，用于窗户、板条和地板。这些橡木来源于奥地利和德国北部。在这栋楼中，开敞形式的办公环境在奥地利并不常见，但是建筑师在有意识地创造一个轻快明亮的现代办公环境。这个建筑的建造系统是由考

Efficient construction in wood. How do you create construction processes that make it possible to reduce production costs while at the same time retaining the scope to create unique wooden buildings of the highest quality? Architect Camilla Schlyter travelled to Vorarlberg in Austria to find out.

WHAT IS THE most important factor for the future of wood construction? The two Austrian architects Helmut Dietrich and Herman Kaufmann answer without hesitation:

"Creating high-quality buildings through a high degree of prefabrication."

To establish industrial construction in wood where quality and economy are the core focus on every front, all the links in the chain need to have the same ambition and drive for quality, in their view.

"We try to achieve a consensus on everything from how the timber grades link up with the forestry and what logs can be used for what part of the building, to how we make use of traditional knowledge in building wooden houses. That requires combining high-tech knowledge and new research with traditional local crafts, as championed by architects and craftspeople, and with the established industrial processes of joiners and construction companies," says Herman Kaufmann. Helmut Dietrich continues:

"In Austria, it's three or four percent more expensive to build in wood than in concrete. But when you factor in the reduced energy consumption and shorter construction times, it's often cheaper in the long run."

Clients therefore need to be clearer about the arguments in favour of a wooden structure. Dietrich, for example, feels that the energy it takes to produce the construction material should be taken into account. In Sweden, work is under way to encourage clients and the construction industry to look at the whole life cycle of the most common materials used. The aim is that, in the near future, consideration will be given to such factors as carbon emissions, energy consumption in production, emissions from transport and the ability to recycle the construction material.

Kaufmann describes how a far-reaching partnership between the client, the developer, the architect, the builder and the industry can create a shared knowledge platform, a wood building culture where practical and theoretical knowledge hold equal status. This creates fertile ground for solutions in a dialogue where everyone's voice is heard and respected. Kaufmann and Dietrich see it as important to create standards for wood construction like those that have been developed in the concrete industry, not least to make building in wood even more cost-effective. So how do you design a production system that can be used for multiple projects while retaining the scope for quality and creating unique buildings?

"Most wooden buildings in Austria are built using project-specific details and designs, but the components are manufactured industrially," says Herman Kaufmann.

A PERFECT EXAMPLE of how far you can go in developing high-quality industrial construction in practical terms is Hermann Kaufmann's newly erected Illwerke Centrum in Montafon, Vorarlberg, one of the world's largest office buildings in a wood hybrid design. The building, which projects 120 m out into an artificial lake, has five storeys and was put up in just six and a half weeks. It is a simple and distinct building in both form and scale, reminiscent of a fine piece of furniture rather than a modern office building for a power company. The quality and attention to detail are breathtaking. The facades are untreated wood, since Kaufmann opposes all forms of fireproofing treatment. He argues that untreated wood can be recycled at the end of its life cycle and also ages beautifully.

The characteristic horizontal projections along the facades have practical benefits such as making window cleaning easier, protecting against sunlight and improving fire safety. However, the design is primarily an architectural choice. Instead of using local spruce, as Kaufmann normally would, more exclusive oak has been used for the windows, panelling and flooring. The oak comes from Austria and from southern Germany. The business conducted in the building takes place in an open-plan office layout, something that remains uncommon in Austria, but was chosen consciously here to create a light and modern workplace.

The construction system was developed by Kaufmann in partnership with concrete company the Rhomberg Group and its subsidiary Cree. The Rhomberg Group predicts major changes in the concrete industry as knowledge about the benefits of renewable construction materials grows. It sees the hybrid system as an opportunity to broaden its horizons.

THE INNOVATIVE CONSTRUCTION system used for the floor structure is based on a combination of spruce and concrete. Large parts of the installations are already built into the ceilings at the factory. The system requires no load-bearing partition walls, which makes for highly flexible spaces. This allows for individual layouts and myriad different uses. The structural elements are exposed, meaning that the wood can be seen, smelt and touched. Both the facade elements and the technical system panels can easily be replaced.

The glulam posts in spruce lock the hybrid floor structure in place, while the weight of the floor structure secures the posts to create a solid structural frame. The posts and beams are dimensioned to meet the 90-minute fire resistance requirement. In Austria, stairwells in buildings over four storeys high must be built in concrete. To minimise construction time, the unusually large floor and facade elements were prefabricated in a factory. Great care was taken to find a method for protecting the wood during assembly. The large floor elements were installed first, and then the facade elements and centre posts were fitted in a single day. Then came the next floor layer, so each level could be made watertight quickly, floor by floor. The construction system proved extremely precise. A

test conducted immediately after installation showed that the structure was completely airtight, making it suitable as a passive building.

Austria has a vibrant culture of traditional woodcraft in both the forest industry and the wood construction industry, which is being keenly preserved and developed. There is a widespread belief in wood as a material and its importance for the future. But this has not always been the case. After the Second World War, Austria stopped building in timber, as had been traditional in the region. Modern materials such as concrete, glass and steel were the preferred choice now, even in a domestic context. The traditional local wooden architecture was replaced in the 1950s and 1960s by the standardised "Swiss alpine chalet", which was built in concrete with decorative wooden details on the gables and balconies.

"It was a response to the way houses in the alpine region were supposed to look, according to an international trend driven by the burgeoning tourist industry," explains Helmut Dietrich.

BUT SOMETHING HAPPENED in Vorarlberg in the 1970s, as a few young architects began designing wooden houses for their friends. The houses had an updated vernacular style and the friends, who often didn't have much money, tended to build the homes themselves with the help of local carpenters. Mayors in the region countered this movement by tightening the planning process. They called the new buildings "stables" and this sort of architecture was definitely not part of their vision. The desire was to move away from the agrarian society with its lowly status, to a high-status urban society. The simple houses sparked a great deal of anger, but the architects stuck to their convictions. Despite the opposition, they sought help from the few surviving local sawmills, joiner's workshops and craftspeople, and together they managed to turn the tide. Once it could be shown that it was possible to build good housing at a low price, the wood building sceptics changed their tune.

The trend in favour of wood accelerated as the green wave swept through the region and local villages began expanding. Architectural competitions were organised for residential developments, schools and other public buildings. All this paved the way for a new generation of architects in the 1980s, including Hermann Kaufmann. Modern designs in wood started to achieve great success in the competitions. A culture of high-quality wooden architecture was created and began to flourish, and now the region leads the world in combining craftsmanship and industrial construction in wood.

Vorarlberg has a keen sense of quality and maintains a close partnership all along the chain from forestry and design to manufacture and construction. According to Kaufmann, there is a strong tradition of investing in architecture, people are proud of the region's craftsmanship, which they feel represents care for the

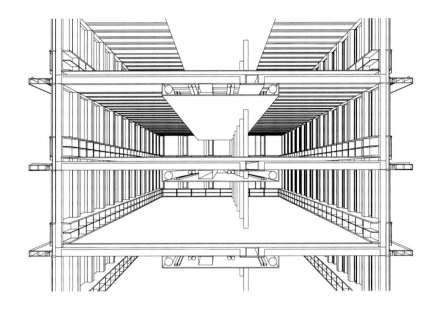

夫曼（Kaufmann）和 The Rhomberg Group 及它们的子公司 Cree 同时开发的。由于更多地了解到再生材料的知识，The Rhomberg Group 预测到了混凝土工业即将面临的变革，所以他们把这个混合系统看为一个拓展领域的机会。

这种创新的建筑系统已经运用于地板构造，这是一种云杉和水泥的混合体。装置中的大部件都已经在工厂里被嵌入了天花板里。这个系统不需要任何承重的隔墙，创造了很灵活的空间。这实现了个人的修改布置和无数其他使用方法。建筑的结构都是裸露的，也就是说人们可以看得见、闻得到、摸得着木头。外墙的布置和技术系统面板都能被轻易地重置。柱子是以云杉为原料的胶合层积材，它们将这个混合地板结构固定，同时，地板的重量确保这些柱子形成一个稳固的结构框架。所有支柱和梁都满足 90 分钟的耐火性测试。在奥地利，超过四层楼的建筑楼梯必须用混凝土制造。为了节省施工时间，尺寸较大的地板和立面元素都在工厂里预制。建筑师们下了很大的工夫来寻找一个能在组装过程中保护木材的方法。大的地板组件先被安装，然后立面部分和中央的支柱都在一天之内安装完毕。紧接着是下一个楼层，这样每一层楼都可以紧凑地建好。实践证明，这个建筑系统是非常严谨精确的。安装之后的实验表明，这个建筑物是完全密封的，使得它适合作为被动式建筑。

在奥地利的林业和木造业领域里，鲜活的传统木工艺文化，仍被热心地保存和发扬。对于木材作为一种原材料在未来的重要性，有着广泛的共识，但是一直以来并不总是这样的。在二战之后，奥地利禁止建筑中木材的使用，虽然那是当地一直的传统。一些现代的材料，比如水泥、玻璃和钢铁成为了更被青睐的材料，甚至是在家里。在 1950 和 1960 年代间，当地传统的木建筑被"瑞士阿尔卑斯山的小屋"完全取代了，那是一种混凝土制造并带有装饰性的木材细节，比如在山形墙与阳台的部分。"那是在不断扩大的旅游业的驱动下形成一种国际化的潮流，那是一种人们以为高山区域房屋应该长成什么样子的回答。"Helmut Dietrich 解释道。但是在 1970 年间，Vorarlberg 发生了一些事件，几位年轻的建筑师为他们的朋友设计木构房屋。这些房屋的设计是经过提升的地方性风格，而且这些朋友们常常并不是很富裕，所以会在当地木工的帮助下亲手建造房屋。当地的市长强调规范了规划步骤。他们把这些新建筑物称为"马厩"，因为这些建筑物显然并不在他们的远景之中。他们的远景是远离低级的农耕社会，走向高级的城市化社会。这些简约的小房子招来了许多指责，但是建筑师们仍旧坚持着信念。顶着反对的压力，他们向当地为数不多仅存的锯木厂、木工作坊和工匠求助，并终于逆转了局势。在他们证明了低价仍能建造一座好的房屋的可能性之后，对木建筑的怀疑者也改变了论调。

随着绿化概念在这个地区的普及，对于木材使用的风潮更被加速推广。他们为居民住宅开发、学校和其他公共建筑举行建筑设计比赛。这为 1980 年代新一代的建筑师们铺路，其中就包括了 Hermann Kaufmann。现代木建筑设计开始在各类比赛里取得成功。一股高质量木构建筑物的潮流文化开始繁荣，而现在这个地区已是世界上将手工工业与工业化生产结合的引领者。福拉尔贝格州对质量有着敏锐的认知，并和林业、制造业和建造业的全产业都形成了密切的伙伴关系。考夫曼认为，这里有着较强的投资建筑的传统，人们为地方性的手工技艺而感到自豪，因为他们认为那是代表了从更长远的角度对于景观和人类的关怀。从社会、经济和文化的角度，这种建筑文化与现在对于可持续性以及地方性资源利用的讨论是正合拍的。那些对于通过严格的标准化解决方案进行预制化建筑的兴趣，其实是基于花费要在设计和木材质量之前先考量的这种想法，可以说是不被称赞的。这样的建筑也被认为是廉价和没有灵魂的。对于瑞典木业的倡导者们而言，这几乎就是一场梦。我们怎样做才能让瑞典有同样的发展呢？

Anders Rosenkilde，木材处理和家具行业协会（TMF）的技术开发主任，认为我们可以从奥地利学习到很多，但是同时指出瑞典自身在这方面已经积攒了很多的经验。在瑞典，约百分之九十的低层建筑都是以木材制造的。"瑞典在预制木结构建造方面占据了强有力的位置，特别是在低层建筑领域中，近十年间在高层建筑物中也有很好的发展。这是因为在 1994 年下达了一道对于木结构高层建筑物的禁令，由此引发了对于消防安全、潮湿、声学和静力学的研究和开发。我们同时拥有建造工业化预制低层住宅的悠久的传统，在未来我们仍然可以在这方面努力。"安德斯罗森基尔德集中列举了一些发展机会，鼓励在瑞典建造更优质木的建筑，创造像奥地利一样的喜人的

landscape and people over a longer perspective. The building culture chimes well with today's debate on sustainability and regional use of resources – both socially, economically and culturally. Interest in prefabricated buildings with strictly standard solutions, where cost comes before design and wood quality, is cool to say the least. Such buildings are considered cheap and soulless. For Swedish advocates of wood, it can sound like a dream. What can we do to encourage the same development in Sweden?

ANDERS ROSENKILDE, head of technical development at the wood processing and furniture industry association Trä- och möbelföretagen (TMF), believes we can learn a lot from the situation in Austria, but points out that Sweden has already come a long way. Around 90 percent of all low-rise buildings in the country are now built in wood.

"Sweden currently has a very strong position in prefabricated wood construction, particularly when it comes to low-rises, but over the past ten years or so also in the construction of high-rise buildings. This is because in 1994 a ban on building high-rises in wood was lifted and that has generated research and development in fields such as fire safety, damp, acoustics and statics. We also have a long tradition of industrially prefabricated low-rises that we can build on for the future."

Anders Rosenkilde highlights a few development opportunities that could encourage construction of more high-quality buildings in Sweden, and create the same positive development as in Austria.

"Swedish architects and structural engineers could be better at the architectural advancement of large wooden buildings. There has been a strong focus on ensuring the designs meet all the technical requirements, but now the technology for building wooden high-rises has fallen into place. There are plenty of good examples of tall wooden buildings and architects are indicating a growing curiosity. Developers are also showing an interest, but at the same time there is a certain amount of caution, since many of them have never commissioned a large project in wood before. Every completed construction project generates positive reactions and experience among the architects, structural engineers, builders and developers," says Anders Rosenkilde.

Susanne Rudenstam is head of the Swedish Wood Building Council. She sees a bright future for the development of prefabricated multi-storey buildings in wood. She also believes that Sweden's commitment to drastically reduce carbon emissions by the year 2050 will provide an extra spur.

"I see a positive spiral ahead. With all

of us, not least the politicians, striving to achieve the climate objectives, more and more clients will be choosing wood for their newbuild projects – municipalities and county councils in particular. The increase in orders will lead to higher demands on the wood industry, and they will become even more skilled at prefabricated high-rise buildings that are assembled on site," says Susanne Rudenstam.

"To keep up with these developments, we need to improve our expertise on all fronts. Universities, colleges and upper secondary schools with a focus on architecture and construction must provide their students with even better knowledge of wood construction. And the wood industry needs to learn from the car industry when it comes to efficient flows, where a focus on learning leads to continuous improvement. If we can succeed in this, then the future is bright for prefabricated and large-scale construction in wood."

ARCHITECTURE & CONSTRUCTION
Hermann Kaufmann has been part of the Vorarlberg architecture scene since 1983 and his practice Architekten Hermann Kaufmann ZT GmbH now has 24 employees in Schwarzach and Munich. They create references to the locality via nature, other buildings and the local community. They focus on sustainable building in general and modern wood construction in particular.

Dietrich/Untertrifaller Architekten was founded in 1994 by Helmut Dietrich and Much Untertrifaller. The practice currently employs 35 people and has offices in Bregenz, Vienna and St. Gallen in Switzerland.

The hybrid beams, a combination of wood and concrete, were developed by Kaufmann in partnership with concrete company the Rhomberg Group and its subsidiary Cree.

European spruce, picea abies, is called white spruce in Austria and grows naturally here. There is no difference in colour between the sapwood and the heartwood, and the colours vary from almost white to a pale yellow-brown.

ILLWERKE ZENTRUM MONTAFON
Illwerke Zentrum was completed in 2013 and constructed from prefabricated elements. The unique feature is the combination of wood and concrete in the floor structure. All the components are on a grand scale, with the facade elements measuring around 12 metres long by 3.2 metres high. All the wood is untreated, so that it can be recycled. The building is the first on this scale to be classed as a "Green Building" in Vorarlberg.

Architect: Hermann Kaufmann for Architekten Hermann Kaufmann ZT GmbH, Schwarzach
Client: Illwerke VKV
Cost: approx. EUR 30 million including planning, design and construction

发展。"瑞典的建筑师和结构工程师大型木建筑方面已经先人一步。曾经重点都放在确保设计能够满足所有技术要求，但现在建造高层木建筑的技术已经很成熟了。有许多高层木建筑的良好范例，建筑师也表现出了更多的好奇心。开发商也表示了兴趣，但同时又保持一定谨慎的态度，因为许多人以前从未委托过大型的木项目。现在，每个完成的项目都产生了积极的反响，同时建筑师、结构工程师、建造者和开发商之间也获得了经验。"Anders Rosenkilde 安德斯罗森基尔德说。

　　Susanne Rudenstam 是瑞典木建筑协会的主席。她对预制多层木建筑的光明未来充满了信心。她同时也相信瑞典对于在 2050 年之前减少碳排放量的承诺会对此有显著的帮助。

　　"我看到一个积极的、螺旋上升的未来。我们所有人，不仅仅是政治家，为了实现气候相关的目标，越来越多的客户在他们的新项目中会选择木材——特别像市政府、县委员会。订单的增加将导致对木材行业更高的要求，他们将变得更善于在现场组装预制的高层建筑。"Susanne Rudenstam 说。"为了跟上这些发展，我们需要提高各方面的专业知识。大学、学院和高中的在建筑和建造专业中，必须提供学生更好的木建筑的知识。木材业在进入高效流动的时候需要借鉴汽车行业，注重学习获得持续进步。如果我们能做到这一点，那么木材的预制和大规模建造就会有光明的未来。"

环抱木材的幼儿园
Preschools embrace wood

Text: Leo Gullbring | 校译：韩佳纹
摄影：Adam Mørk, Tim Crocker, Bernardo Bader Architects

| Råå in Helsingborg, Sweden by Dorte Mandrup architects
| Published 16 June 2014

The ecocycle and children's dreams. Wood is more than an important ingredient in environmentally and socially sustainable building. The material also has a uniquely warm look and tactile properties that appeal to children. Let us take you to three different and vibrant Preschools, where the focus has very much been on ecolabelled materials and the dreams and needs of the children.

BUILDING FOR CHILDREN is not easy. Their energy explodes all the boundaries, since they are not yet familiar with the diktat of adult life that you have to spend most of your day sitting still. It is therefore a tough challenge to create architecture for children, something that makes architects self-critically question regular architecture.

环抱木材的幼儿园

Råå Preschools in Helsingborg, right by the waters of Öresund, looks more like a sculpture than a kindergarten. The school has an unbroken geometric shape with a facade of narrow vertical robinia cladding. Not exactly what you might expect alongside the traditional brick facades of the old school buildings. The interior is at least as fascinating with all its angles and nooks, large windows and triangular lanterns that illuminate the open environment. In an uninsulated side building, the little ones can sleep off their lunch in little wooden wagons, indoors and yet outdoors.

"The key is to encourage the children's motor skills with spaces where they can climb and negotiate steps," says Pernille Svendsen, architect at Danish practice Dorte Mandrup, which is responsible for this project.

"Wood is a fantastic material that brings warmth and has a rare materiality. We've used veneered panelling in the design, a natural material that looks almost like we found it on the beach."

Preschools embrace wood

关于孩子们的梦想与生态循环。木材一直是环境与社会的可持续发展建设中相当重要的元素。木材还具有独特温馨的外观和触感，从而十分吸引着儿童。现在让我们带你到三个不同的、充满活力的幼儿园。它们一直以来就因为其材料所具有的生态标签和对孩子们的梦想和需求的满足而备受关注。

为孩子们盖房子并不容易。他们到哪儿都如此充满活力，如果他们还和你不熟就不会听从成年人的命令，因此你不得不花费大部分的日子坐在那观察他们。建筑师也不得不对他们原先的常规构架做出自我批判，这也是为什么为孩子们创造一个建筑是如此巨大的挑战。

Råå Preschools 坐落于赫尔辛堡，厄勒海峡的东面，它看起来更像一个雕塑而不像幼儿园。这所学校有着不间断的几何形状，外立面由狭窄且垂直的槐木覆层构成。与你所期望的那种紧挨砖砌外墙的老校舍的建筑不同，它的内部无论是从哪个角度看哪个角落都同样引人入胜。开放环境的采光来自于大面积的开窗和富有创意的三角灯笼。在侧楼，这些小家伙们可以在午餐后睡在他们的小木头汽车里、室内、当然还有室外。

负责这个项目的丹麦建筑公司 DorteMandrup 的建筑师 PernilleSvendsen 说："为了启发孩子们的运动技能设计可以攀爬和来回活动的台阶空间。"

"木材是一个可以带来温暖的绝妙材料，它有一种极为特殊的材料性，我们在设计中使用了一种天然的看起来几乎像是在沙滩上找到的胶合镶板。"

Pernille 一直被建筑承包商是如何难以掌握几何的问题所困扰，这同样也是她和她的同事同样需要解决的问题。一个出现在建筑师方案中的 1：50 的模型对主承包商来说无疑是个惊喜，当然对于木工和钳工来说也是一个巨大的帮助。

"我们用过参数化设计吗？没有，我们在 2D 平面工作。"PernilleSvendsen 笑着说，"它是如此难以置信的复杂以至于我们不得不在办公室里一个接着一个的做模型。"

比起瑞典的建筑风格，在建筑行业中建筑师 Dorte Mandrup 和她的丹麦同事代表了更具概念性的建筑风格，主要因为建筑师在行业中更加独立的地位，也因为他们在艺术学校里学习而不是在技术学校。对于 Råå 幼儿园的设计工作是非常直接的儿童空间为起点的，是一种完全不同于成年人的空间。当他们一进来的时候，就会到处乱跑乱爬，随意进入不同种类空间，这就导致对整个空间的设计无须具有过度的完成性、指令性。因此他们有一个可以供他们任意想象玩耍的出口，但是 PernilleSvendsen 强调他们也需要一些私密空间让他们可以和最好的朋友待在一起。而且随着他们即将在学校里学习超过十年的时间，为他们创造出可以激发他们创造力的环境也尤为重要。

在衣帽间一共有两个游戏房，早上孩子们可以在上面的一个游戏房和他们的父母挥手告别。这种设计要归功于孩子们和学校工作人员们在设计的一开始就参与其中了。

环抱木材的幼儿园

PERNILLE RELATES HOW the contractor had difficulty getting to grips with the geometry, something that she and her colleagues also had to wrestle with. A 1:50 scale model appearing on the architect's bill was a surprise for the main contractor Peab, and a great help to the carpenters and fitters.

"Did we use parametric design? No, we worked in 2D," says Pernille Svendsen, laughing. "It was so incredibly complex that we had to build model after model here at the office."

The architects at Dorte Mandrup and their Danish colleagues in the industry represent a much more conceptual architecture than Sweden is used to seeing. This is due in part to a more independent role for the architects, but also to the fact that architects are trained at art schools, not technical colleges. The starting point for work on the Preschools in Råå was that Preschools children are incredibly direct and occupy a space differently to adults. As soon as they come in, they start running around and climbing. They need access to many different types of space, which need to be not too finished and prescriptive, so they have an outlet for their imaginative play. But they also need more private spaces where they can be 'alone together' with their best friend. And as they are going to be learning at school for more than 10 years, it is important to create environments that stimulate creativity through experiences and challenges, argues Pernille Svendsen.

There are two playhouses in the cloakroom, and up in the top one children can wave goodbye to their parents in the morning. This has come about because pupils and staff were involved in the design right from the outset.

"They told us what their dreams are and we made use of our experience on similar projects in Denmark. Light plays a particularly important role, and of course the opportunity to play. It's important to scale everything properly for children," says Pernille Svendsen.

WHEN IT CAME TO building Hilden Grange Preparatory School in Tonbridge, in the English county of Kent, involving the children was not as easy. The parents and staff, on the other hand,

环抱木材的幼儿园

were enthusiastic about the tired temporary classrooms finally being replaced with a modern new complement to the old manorial school building up on the hill.

"We held several workshops with all the youngest children about how we would build an amphitheatre with an unusual and challenging outdoor space," relates architect Roger Hawkins, explaining that in other projects he is able to involve the pupils a little more, but in this case the school itself was not the client. That was the Alpha Plus Group, which manages the school.

"This is a sensitive site and we wanted to create something that stood in stark contrast to the red brick architecture of the century-old main building. We wanted something softer and more welcoming for the children, who are aged 4 to 13, more like garden pavilions that sit comfortably in the landscape. We brought in a landscape architect, who designed the wildflower meadow on the roof of the east wing," says Roger Hawkins.

The two wings of the school are joined together by a school hall and canteen, which can accommodate all the school's pupils. Roger sees wood as an obvious choice. Like the client, he saw wood as a way of speeding up the building process, and as a cost-effective solution, not to mention the purely environmental benefits of wood, benefits that were spelled out in life cycle analyses. To avoid too polished a look for the exposed design, cross laminated timber was chosen with traces of knots and irregularities.

"The choice of a prefabricated design using CLT made the most sense. Access to the construction site was not exactly great, along a narrow road that would also be used by the school's staff and parents," says Roger Hawkins.

The school library rises over several floors and strong accents of colour have been used throughout to mark the different parts of the building. The roofs are tiled with Western red cedar shingles, and the facades are larch, which requires minimal maintenance.

"The structural timber frame for the central section is a hybrid, where we used steel struts to brace the long spans that measure between 15 and 20 metres. We've managed with natural ventilation, except in the canteen, kitchen and toilets. Overall, this has been a highly cost-effective project," states the architect.

Roger Hawkins believes the UK has a long way to go to

Preschools embrace wood

PernilleSvendsen 说:"他们告诉了我们他们所梦想的然后加上我们在丹麦做的相似项目的经验。我们总结认为光线扮演着一个相当重要的角色,特别对于提供玩的空间至关重要。同时,设计每样东西还要考虑到适用于孩子们的尺寸。"

当建造英国肯特郡汤布里奇的 Hilden Grange 幼儿园时,涉及到孩子们的事情就不那么容易了。不过另一方面,家长及工作人员们对于那个建在山上老式庄园式的学校的老旧临时教室最终被一个现代化的教室所取代表示热情欢迎。

"我们和所有年轻的孩子们就如何建造一个与众不同的且充满挑战的户外空间举办了好几次专门的研讨会。"建筑师 Roger Hawkins 解释道:"在其他项目里面他可以考虑小学生们多一点,但在这个情况下学校本身并不是客户,而是管理这所学校的 Alpha Plus Group。"

Roger Hawkins 说:"这是一个敏感的地方,我们想建造一个可以和百年来的主流红砖建筑形成鲜明对比的东西。我们希望它更加柔软并且更加受四到十三岁的小朋友们的欢迎,同时也希望它更像是一处置身于美丽风景中可以舒服坐着的花园景观建筑。所以我们请来了一个景观设计师,他设计了一个在屋顶东翼的野花草地。"

学校的两翼各坐落着可以容纳所有学生的一个大礼堂和一个餐厅。Roger 认为木材是一个很显然的选择。和客户一样,他认为木材是加快建设进程的一种方式,是一种成本效益的解决方案,更不用说木材的纯环境效益和在生命周期中被阐明的好处了。为了避免过多的打磨掉暴露式外观设计，带有节点和不规则痕迹的正交胶合木被选用。

Roger Hawkins. 还说："在选择预制设计中对 CLT 的使用是最有意义的。同时施工时，进入学校的施工现场显然是不好的，因此家长和学校工作人员沿着一条狭窄的小路进出学校。"

学校图书馆是个多层建筑，各种强烈的色彩用在整个建筑上以显示该建筑的各个不同区域。屋顶瓦片用的是西部红雪松木瓦，其外墙面是落叶松，这样可以使它需要最少的维护。

"木结构的中央部分是混合结构，我们使用钢支柱支撑15—20 米的大跨度。除了餐厅、厨房和卫生间，我们采用自然通风管理。总的来说，这是一个极具成本效益的项目。"建筑师陈述道。

Roger Hawkins 认为英国在木结构建筑方面还有很长一段路要走才能达到其他欧洲国家的水平。最大的问题是要找到有勇气按照非传统理念思考的客户。

在奥地利找到一个想要木结构建筑的客户根本不是问题。这个国家有令人印象深刻的现代木结构建筑群，至少在奥地利西部福拉尔贝格州。在这里的建筑都有一种很罕见的细节

reach the same level as other European countries, not least Sweden, when it comes to building in wood. The big problem is finding clients with the courage to think along non-traditional lines.

FINDING CLIENTS WHO want a wooden building is rarely a problem in Austria. The country has an impressive portfolio of modern wooden architecture, not least in Vorarlberg in western Austria. Here there is an all too rare sense of detail in the architecture, largely thanks to recurring competitions for public projects. The young architect Bernardo Bader, who runs the architectural practice of the same name, is one of the figures driving the country's wood building traditions forward, as epitomised in the Kindergarten Susi Weigel in the rural town of Bludenz. The Preschools has an almost graphic simplicity that is reinforced by the beautifully detailed fixtures and fittings in wood, which focus on children at play.

"Our aim was to work with natural and strictly organic materials," explains Bernardo. "Our client, the town of Bludenz and its mayor Josef Katzenmayer, contributed timber from its own forest, which cut our construction costs and made the project more sustainable and rooted in the locality. For the facade, we used Scots pine that will age naturally without the need for maintenance. We avoided all non-natural building materials, and all the wood is treated with natural products."

Framing the views are generous windows with sills that make lovely window seats, while the locally harvested timber ties in with the surrounding landscape. Interior walls in silver spruce give a light and airy environment that is accentuated by the clean, open layout, which links together the two floors with a freestanding staircase and mezzanine. The building is frank in its self-confidence. The exterior acts as a geometrically precise skin for its inner workings. The concrete in the load-bearing walls in the middle of the building is as boldly exposed as the different types of wood in the facades, walls, ceilings and floors.

"Most of us have children, and my wife is a Preschools teacher. I spent an awful lot of time studying the way my children play outside and inside to try and understand what they like and what they miss, so I could incorporate that into our architectural design.

All the furniture is on a slightly smaller scale, of course, and all the spaces are multifunctional."

CHILDREN'S AUTHOR SUSI WEIGEL has lent her name to the Preschools, and her illustrations inspired colour specialist Monika Heiss to paint the thin stripes that line all the doors like a grid. The school's five classes each have their own playroom as well as a more intimate space. In addition, all the 100 children share a large area for joint activities.

"The Preschools is located on the edge of the town, and we've chosen to have the large windows facing the landscape, so the children can see the mountains and fields instead of other buildings. The children feel more like part of nature than part of the town."

With its restrained look, Bernardo Bader's school building would not be out of place in a spa complex. Nevertheless, the children feel perfectly free to use the building exactly how they want to.

THREE Preschools & THREE ARCHITECTS
Råå Preschools
Helsingborg, Sweden
The Preschools was completed in August 2013. Danish architectural practice Dorte Mandrup is behind this design, and many other lauded projects, primarily in Copenhagen. Their most recent new commissions in Sweden are for IKEA's new offices in Malmö and the School of Economics & Management at Lund University.
Architect: Dorte Mandrup
Client: City of Helsingborg
Structural engineers: Tyréns and Ramböll
Cost: SEK 85 million (price includes an extensive remodelling, renovation and damp proofing of Råå Södra Skola).

The Ritblat Building, Hilden Grange Preparatory School
Tonbridge, Kent, UK
Hilden Grange was completed in 2012 to a design by Hawkins\Brown, a practice founded by Roger Hawkins and Russell Brown in 1998, which now employees 100 people in Clerkenwell. Their accolades include Sustainable Practice of the Year.
Architect: Hawkins\Brown
Client: Alpha Plus Group
Structural engineer: Price & Myers
Cost: Approx. SEK 45 million

Kindergarten Susi Weigel
Bludenz, Vorarlberg, Austria
Kindergarten Susi Weigel was completed in 2013. The building was designed by architect Bernardo Bader, who launched his own practice in the early 1990s. He works with extremely precise and exacting architecture that integrates with the landscape, and his work has won him the Aga Khan Award.
Architect: Bernardo Bader
Client: Amt der Stadt Bludenz
Structural engineer: Brugger Ingenieure
Cost: not available

感，这主要归功于经常会有的公共项目竞赛。年轻的建筑师 Bernardo Bader，领导着一个有着相同名字的建筑设计事务所，同样也是推动这个国家传统木结构建筑发展事业的人之一，其代表建筑是坐落于苏西威格尔·布鲁登茨乡镇的一个幼儿园。这所幼儿园有一个极简图形化的建筑外观，而且由十分精美而细致的木结构节点连接件加固，其主要目的就是让孩子们可以尽情玩耍。

Preschools embrace wood

"我们希望能完全地利用天然有机材料。"Bernardo 解释道："我们的客户是 Bludenz 市和它的市长 Josef Katzenmayer，并提供了当地的森林用作项目的木材，这大大减少了我们建设开支并且使得整个项目更加可以可持续性的根植于当地。在建筑物的正立面，我们使用苏格兰松从而让其可以自然老化而不需要维护。我们没有使用任何非天然建筑材料，所有的木材都是经过纯天然加工处理的。"

特别宽大的窗户下的窗台正好可以作为可爱的靠窗座位供人在此欣赏美景，同时取自当地的木材也使建筑更加紧密的融入周围的环境。内部墙面用银云杉装饰，并且利用光线和通风环境，强调干净、开放式的布局，同时独立式的楼梯将上下两层和夹层连接起来。这栋建筑用一种很直接的方式表现出了自信。它在外部用精确的几何造型来表现其内部构造。建筑中间承重墙上的混凝土就像立面上、墙上、天花板上和地板上的木材一样大胆的裸露着。

"我们中的大多数人都有孩子，而且我的妻子也是一个幼儿园老师。我花了很多时间研究我的孩子们在外面玩耍的方式，试图了解他们喜欢什么，以及他们在想什么，这样我可以把这些融入到我们的建筑设计中去。所有的家具的尺寸都被稍微改的小了一点，当然，所有的空间都被设计成是多功能的。"

儿童作家 SUSI WEIGEL 将她的名字借给了学校，并且她的图书插图给了色彩学家 Monika Heiss 用细条纹将门涂的像网格一样的启发。学校的五个班级都有他们各自的活动教室和一个更加私密的空间。此外，还有一个更大的空间可以供所有的一百多位孩子们共同活动。

学校坐落于小镇的边缘，因此我们选择设计了面向自然景观的大窗户，这样孩子们可以看到群山和田野而不是其他建筑物。相对于城镇来讲，孩子们将觉得自己更是大自然的一部分。

建筑的外观设计受到了严格的控制。尽管如此，孩子们仍然可以完全自由的以他们想要的方式使用这座建筑。

THREE Preschools & THREE ARCHITECTS（三个幼儿园 & 三名建筑师）
Råå Preschools
赫尔辛堡，瑞典

该幼儿园于 2013 年 8 月完工，负责这次设计的是丹麦的建筑事务所 DorteMandrup。他们还有很多值得称赞的项目作品，不过大部分都在哥本哈根。最近，他们在瑞典参与设计的项目包括宜家在 Malmö 的办公室设计和隆德大学的经济与管理学院建设。

建筑师：DorteMandrup
客户：City of Helsingborg
结构工程师：Tyréns and Ramböll
总造价：SEK 85 million (price includes an extensive remodelling, renovation and damp proofing of RååSödraSkola).

The Ritblat Building, Hilden Grange Preparatory School（利特布莱特建筑，希尔顿·格兰奇幼儿园）
英国肯特郡汤布里奇

Hilden Grange 于 2012 年完工。负责项目设计的是 1998 年由 Roger Hawkins and Russell Brown 共同建立的 Hawkins\Brown 设计事务所。如今，该事务所在克拉肯韦尔有超过 100 名员工。他们的荣誉包括年度可持续设计奖。

建筑师：Hawkins\Brown
客户：Alpha Plus Group
结构工程师：Price & Myers
总造价：Approx. SEK 45 million

Kindergarten Susi Weigel（苏西·魏格尔幼儿园）
奥地利福拉尔贝格州 bludenz

  苏西·魏格尔幼儿园于 2013 年建造完成。项目由上世纪九十年代早起就拥有了自己的设计事务所的建筑师 Bernardo Bader 。他以极其精确和严格的体系结构设计与景观设计相结合，使他的作品赢得了阿迦汗奖。

建筑师：Bernardo Bader
客户：Amt der StadtBludenz
结构工程师：BruggerIngenieure
总造价：not available

Preschools embrace wood

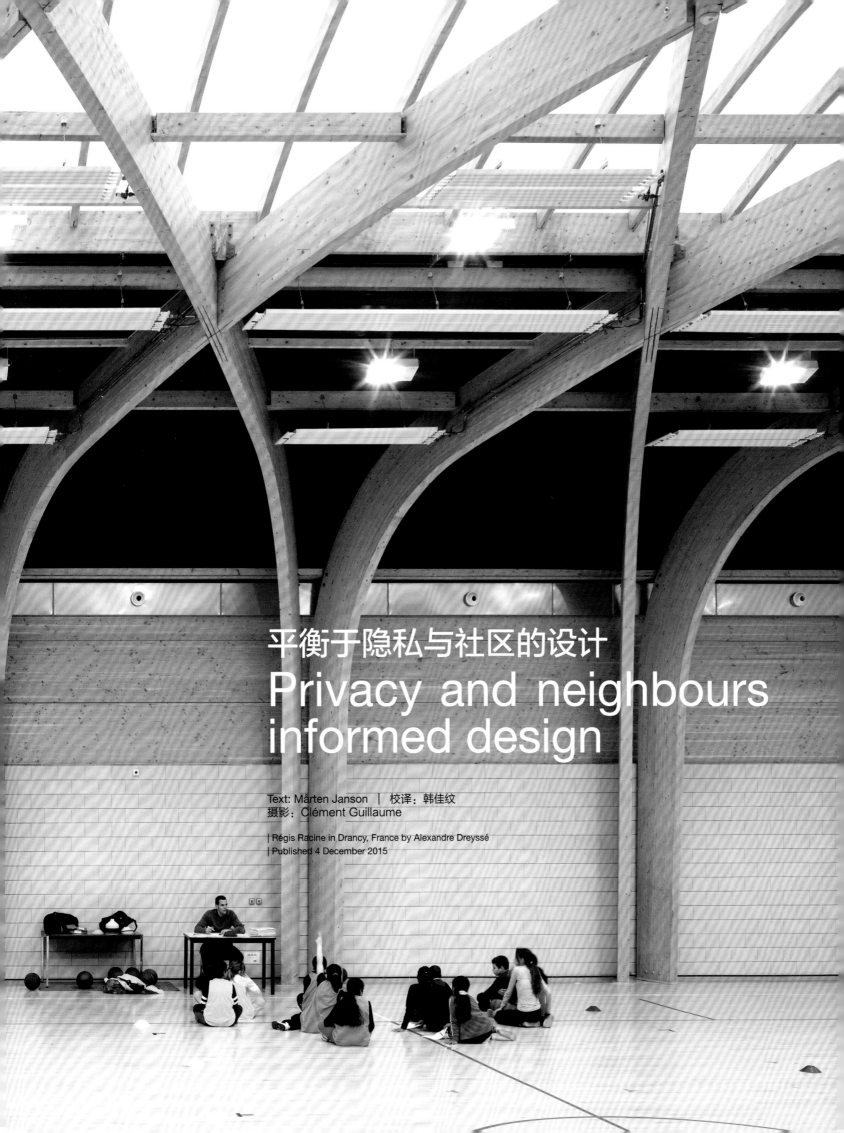

平衡于隐私与社区的设计
Privacy and neighbours informed design

Text: Mårten Janson | 校译：韩佳纹
摄影：Clément Guillaume

| Régis Racine in Drancy, France by Alexandre Dreyssé
| Published 4 December 2015

Regis Racine 体育馆为破烂不堪的巴黎郊区带来新的尊严。
灵感来自于几世纪前的有顶棚的市场，和瑞典功能主义的理念。
最后成果为只用胶合梁连接交错而成的教堂式的拱形屋顶。

德朗西是一个深受社会问题困扰的巴黎郊区，其历史也使它声名狼藉：二战期间，德国侵略军在这里建立了拘留营，大部分住房存量由低层建筑组成，曾经的农田则被用来盖这些建筑。当地市政府花了几年时间计划建立一个新体育馆，但都以失败告终。直到 Alexandre Dreysse 和他的同事们接受这项任务，这一计划才得以实现。他们想赋予这个被忽视的郊区一种新生——种新的自尊。

平衡于隐私与
社区的设计
Privacy and
neighbours
informed design

Régis Racine gymnasium brings new dignity to a down-at-heel suburb of Paris.
Inspiration comes from the covered markets of previous centuries and from legends of Swedish Functionalism. The result was a cathedral where only glulam beams would do.

DRANCY IS A suburb of Paris plagued with social problems. The area is also stigmatised by its own history: during the Second World War, the German occupying forces set up an internment camp here. The housing stock largely comprises low-rise buildings built on former farmland. The municipality had spent years planning for a new gymnasium, but nothing ever came of it. When the task of realising the plans finally went to architect Alexandre Dreyssé and his colleagues, they wanted to help give the neglected suburb a new start – and a new sense of dignity.

Alexandre Dreyssé was born in Strasbourg and comes from a family of architects. He is the fourth generation to go into the same profession. His father, Henri Dreyssé, made a name for himself in the 1980s with a number of large wooden structures, including two residential buildings in Alsace. Alexandre Dreyssé trained in Paris, where for a while he also worked for the Cuban architect Ricardo Porro, before setting up his own practice. Dreyssé took his inspiration for the gymnasium, named after French basketball player Régis Racine, in part from the covered markets of the 19th century such as La Halle Secrétan and Baltard in Paris. But there are also influences from 20th-century Swedish architects Sigurd Lewerentz and Ralph Erskine. Another inspiration for the project was the work of British architect Edward Cullinan.

TODAY RÉGIS RACINE is used primarily by local schools and sports clubs. The sports hall itself is optimised for regional basketball tournaments, but the building also houses a studio for ballet practice. The impressive vaulting and seven metre ceiling height make it hard not to think of a Gothic cathedral – even if the arches are glulam beams rather than sandstone.

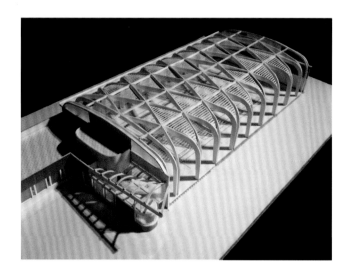

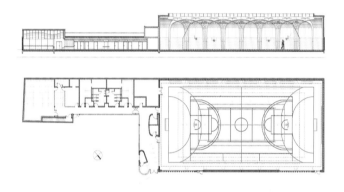

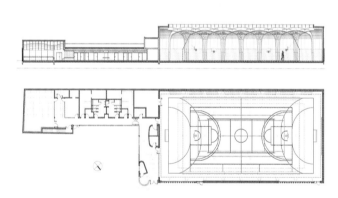

Alexandre Dreysse 出生于斯特拉斯堡一个建筑世家，他已经是第四代从事建筑行业，他的父亲 Henri Dreysse，在 80 年代以他大量的大型木结构建筑而闻名，其中包括阿尔萨斯的两处住宅建筑。Alexandre Dreysse 受训于巴黎，并在组建自己的工作室之前，曾为古巴建筑师 Ricardo Porro 工作过一段时间。

Dreysse 将自己的灵感注入这座体育馆，并以法国篮球运动员 Regis Racine 的名字为其命名。其一部分灵感来自于 19 世纪的有顶棚的市场，如巴黎的 La Halle Secretan 和 Baltard。但同时也受到 20 世纪瑞典功能主义建筑师

平衡于隐私与社区的设计
Privacy and neighbours informed design

平衡于隐私与社区的设计
Privacy and neighbours informed design

There are several reasons why they chose wood, according to Alexandre Dreyssé. Firstly, there was the huge roof span: with an area of 44 by 22 metres, glulam beams were the only reasonable option. Building in prefabricated wooden modules also brought less inconvenience to the neighbours in a densely populated area. And when the client, the local municipality, expressed its desire for the building to use a durable natural material, that really sealed the deal. In terms of the building's appearance, the character of the site also guided the architects — not least the fact that the façade is almost entirely surrounded by other buildings. "Getting daylight into the building was a challenge," says Alexandre Dreyssé.

The solution was opaque glazing integrated into the roof structure. To cut down on weight, polycarbonate plastic with insulating air channels was used. It was also easy to adapt to the arched shape of the roof. The fact that polycarbonate plastic is cheap was a bonus. "A simple and honest solution, a bit like a potting shed!"

The parts of the façade that are visible contain double-glazing. The material provides sufficient insulation most of the time, but when it gets really hot outside, it can get a bit warm inside too, explains Alexandre Dreyssé. And since only a small section of the façade is visible from the outside, the main focus was on the interior. "The building became kind of inverted – we invested in high quality on the inside!"

The overall design then emerged from studying a number of simple paper models.

"We didn't really have a particular vision of what we wanted to do, instead we looked at what was appropriate for the site. The vaulted ceiling came about because we didn't want to build too high a structure next to the neighbouring low-rise housing."

THE COMBINATION OF WOOD AND BLOCKWORK has become something of a signature for Dreyssé's practice. But again, the geographical location of the plot came into play. It faces a number of other properties – erecting scaffolding on the outside would have required negotiations with a long list of property owners. Construction started with a cast concrete slab

Sigurd Lewerwntz 和 Ralph Erskine 的影响。另一部分灵感则来自于英国设计师 Edward Cullinan 的作品。

现在的 Regis Racine 体育馆主要提供给当地学校和运动协会使用。运动大厅针对篮球进行了优化,以用于地区比赛,同时设有一个工作室供芭蕾舞训练。其令人印象深刻的拱形结构和 7 米高的天花板,即使是使用胶合木而不是砂岩建成,也无法不使人联想到哥特式大教堂。

Alexandre Dreysse 认为选择木材有几个原因。首先,屋顶的跨度是巨大的,面积有 44x22 平方米,胶合木是唯一合理的选择。用预制木材部件建造也为人口稠密地区的邻里间减少了不便。而当作为客户的当地政府提出想为体育馆选取一种耐用的天然材料,这个选择才真正定下来。依据建筑物的外观,项目选址的特殊性也影响着建筑师,尤其是建筑物正面几乎完全被其他建筑所环绕这一实情。

"如何让阳光照进建筑物内是我们面临的一大难题。"Alexandre Dreyssé 说道。

最后解决方法是将不透明玻璃结合进顶部结构中。为减少重量,使用空气隔热层的聚碳酸酯塑料,这种材料也更容易适配屋顶的拱状。事实上,聚碳酸酯塑料成本低也是额外一大优点。

"这是一个简单又实在的方法,效果有点像盆栽棚。"

建筑物的部分表皮含有双层玻璃。这种材料在大部分情况下都可以保持充分隔热,但如果外部温度实在很高时,建筑物内部可能也会有点热。Alexandre Dreyssé 解释到,同时,因为外观只有一小部分是从外可见的,所以主要的关注点都放在了内部。

"这座建筑可以说是反转性的创建了一个高质量的内部空间。"

全部的设计,都通过对一些简单的纸质模型的实验和研究。"关于我们想要做的,我们脑海中并没有一个特定的形象,相反,我们更多的是在考虑对于项目地址来说,什么更合适。之所以有拱形顶棚这个想法,也是因为我们不想在周围低层住宅中建一个过高的项目。"

平衡于隐私与社区的设计
Privacy and neighbours informed design

平衡于隐私与社区的设计
Privacy and neighbours informed design

and then the building was erected 'from the inside'. The lower section comprises 20 cm thick, prefabricated concrete blocks that were left untreated on the outside. Once the block walls were up, the builders could carry out all the woodwork within the cordon set up around the site.

ON THE INSIDE, the dividing walls are built from the same concrete blocks as the exterior, although half as thick. Between these are sections of wood using a traditional stud system and an unpretentious cladding of OSB panels.

"Since the building is in a suburb with a history of social problems, we wanted durable materials that could take rough treatment."

The choice of materials such as blocks, wood and plastic helps to create a horizontal impression on the outside. The standing seam metal roof is also extensively punctuated with sections that admit light. By studying the models, the architects found that the ceiling would look much more exciting if the arches were crossed over each other. Most of the daylight comes from above, creating a natural centre inside the sports hall.

"You could say it feels like a cathedral, but it wasn't a look we were intentionally trying for. It was more a consequence of our work with the models," states Alexandre Dreyssé.

"It was actually quite a simple and intuitive process. But structural wood engineers Tec Bois, who specialise in this kind of thing, were also a great help. The engineers showed huge respect for our architectural principles. They were the ones who made the project possible!"

In line with the Functionalist ideals that inspired the project, all the technical installations were left uncovered: electricity, heating, water and waste – everything is completely exposed beneath the ceiling.

"We didn't want to hide things in ducting. It was a choice that also led to a good partnership with all the different trades – we showed that we value their work!" There is also a Functionalist edge to the part of

木材与砖造物的结合，某种程度上可以说是 Alexandre Dreyssé 工作室的标志。但是，项目的地理位置使得这一举动颇难实施。因其正面还有许多其他住宅，如果想要在外面竖立支架，就必须得和大量的住宅所有人协商。工程从现浇混凝土楼板开始展开，于是这座建筑物的建设就从内部开始了。原先计划使用的混凝土预制板则未经处理就被搁置，底段的预制板有 20cm 厚。墙体一经建好，工人们就可以开始将木材运进工地四周拉起的隔离线内。

在建筑物内部，使用了和外部一样的砌块来建隔墙，但厚度仅为外部的一半。木材的部分，则是使用了一种传统柱体系和不加修饰的 OSB 板覆层。

"考虑到建筑所在地是一个有着社会历史问题的郊区，我们需要的材料必须耐用且经得起粗加工。"

在材料选择上，砌块、木材和塑料都有利于制造出一种横向感。立接缝金属屋面也被反复打断引入光线。通过对模型的研究，建筑师们发现，如果拱形之间相互交错穿插，能令天花板看上去更有趣。大部分日光自上而下照进馆内，在大厅形成一个天然的聚焦点。

"你可以说它像一个大教堂，但我们并非故意为之。它更像是我们通过模型得出的结论。" Alexandre Dreyssé 说。

"项目设计本身是一个相当简单且遵循本能的过程。但在木结构工程上，幸好有 Tec Bois 这样的专业木结构工程师，他对我帮助很大。工程师们都最大程度的尊重、配合我们的建筑设计理念。是他们，成就了这个项目。"

与灵感来源之一的功能主义观念相符，所有的技术性的装置都是暴露在外的：电、暖气，水管和垃圾管道——这一切都是完全暴露在天花板下。

"我们不想把事物都用管道隐藏起来，这样的选择也能促进我们与其他各方面负责人的合作关系——我们以此表示对他们工作的重视。"

功能主义的前沿理念还体现在，体育馆从街上看是完全可见的，大胆的曲线形表面使用了道格拉

斯冷杉,也就是众所周知的花旗杉——也正是云杉为胶合梁提供了结构构架。

大型自承重屋顶结构在一辆升降机和吊车的帮助下,每次运一块,才能安置到位。每一片拱形都由四个预制部件构成,部件由阿尔萨斯一家公司制作并运输到施工地。选取法国北部孚日地区的云杉为生产材料。这些部件通过螺栓和直角金属配件固定在一起。而这些嵌固件必须尽可能隐藏。Alexandre Dreyssé 并不知道这些胶合梁、螺栓、聚碳酸酯塑料和金属屋面的实际重量。"尽管知道可能弄明白会更好。"

ATELIER DREYSSÉ

工作室由 4 个员工在 2008 年成立。工作室大部分业务都立根于法国,尽管 Atelier Dreysse 目前正在欧洲参加各种建筑设计竞赛,并即将开始摩洛哥一处别墅的设计,此后还有几处小型黏土建筑项目。

客户:Drancy Commune.
结构工程师:Tec Bois.
总造价:EUR 2,2 million.

the building that is fully visible from the street. The daringly curved surfaces work with the façade material – Douglas fir, or Oregon pine as it is also known – and here again spruce glulam beams provide the structural frame.

THE HUGE, SELF-SUPPORTING ROOF STRUCTURE was put in place with the help of a scissor lift and a crane, one piece at a time. Each arch comprises four prefabricated sections which were made by a company in Alsace and transported to the site. The material is spruce grown in the Vosges region of northern France. The sections are held together with bolts and angled metal fittings. These fixings have been made invisible as far as possible. Alexandre Dreyssé has no idea what this mass of glulam beams, bolts, polycarbonate plastic and metal roofing actually weighs.

"Although it would probably be good to find out."

ATELIER DREYSSÉ
The practice has four employees and was founded in 2008. Most of its business is based in France, although Atelier Dreyssé is currently competing in various architectural competitions in Europe and to design a villa in Morocco. The latter also includes several small clay buildings.

Client: Drancy Commune.
Structural engineers: Tec Bois.
Cost: EUR 2,2 million.

简单的材料体现想象力和对称
Imagination & symmetry come together in material simplicity

Text: Erik Bredhe | 校译：蒋音成
摄影：11h45, Teqtoniques, Renaud Araud

| Paul Chevallier school & Hacine Cherifi sports hall in Rillieux-la-Pape, France by Teqtoniques
| Published 17 March 2016

大胆的想象与简单的对称相遇了。在法国小镇 Rillieux-la-Pape，一所有屋顶花园的学校与一座巨大的体育馆共同形成了一个融入周边环境的教育场所。

在亲近或身处大自然氛围中受教育的孩子更容易接受知识。他们也更冷静，更有创造力，更机警。老师们也能在亲近自然的环境中感到不一样的责任感，这种责任感会依次传递给他们的学生。这是最近刊登在赫芬顿邮报的一篇文章通过广泛研究得出的结论。Paul Chevallier 学校坐落在法国里昂北部的小镇 Rillieux-la-Pape，它在多方面都符合邮报文章中所描述的绿色、有促进性的学校环境。在绿树成荫的 Brosset 公园附近，有两栋令人印象深刻的木建筑，它们就像是从经典奇幻故事指环王中走出来的学校。

这所学校最显著的特征就是起伏的屋顶。屋顶从较高的建筑缓缓延绵至较低的建筑，便将两者连接起来了。屋顶上有草坪和野花，甚至还有一条小路。每个屋顶都有两个平台，供学校里的孩子进行室外课程。在那儿能欣赏到小镇和周围群山及森林的壮丽景色。

"我们希望学校的建筑和大自然能有紧密的联系。铺设了草皮的屋顶，并不是为了给学校近邻提供观赏的，我们希望它们能被师生们使用。"Tectoniques 建筑事务所的 Lucas Jollivet 解释说。

Tectoniques 事务所设计的大型建筑还包括 Hacine Cherifi 体育馆。该项目曾因 2009 年金融危机的来袭而不得不被搁置。欧洲经济恢复后，项目得以继续开展，体育馆随之很快完成。这三栋建筑展示了人们对木材这一今后主要建筑材料的信念。Tectoniques 事务所因其对环境问题、可持续发展问题的严肃立场以及对木材充分的利用在法国享有名声。

"我们热爱木材。我们希望人们看到它！我们并不希望木材被隐藏或者看起来不像木材。混凝土就是混凝土，木材就是木材。理想状况下，材料应同时兼具触感和实用性。体育馆完很好地用了混凝土基础，这样球打到墙上时就不会损坏墙体或留下印记。"Lucas 说。

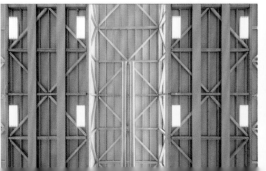

Eye-catching imagination meets simple symmetry. In the French town of Rillieux-la-Pape, a school with flowers and grass on its roof has been joined by an enormous sports hall to create an educational site that merges in with its setting.

CHILDREN WHO ARE taught close to or in nature are more receptive to knowledge. They are also calmer, more creative and alert. Teachers also feel a different commitment in surroundings close to nature, which in turn gets passed on to their pupils. This was the conclusion of extensive studies presented recently in a major article in the Huffington Post. The French town of Rillieux-la-Pape, just north of Lyon, is home to the Paul Chevallier school, which in many ways encapsulates the very green, stimulating school environment that the article describes. Adjacent to leafy Parc Brosset stand a couple of impressive wooden buildings that are as like a school as they are something from the classic fantasy tale Lord of the Rings.

Perhaps the most striking feature of the school is the undulating roof, which slopes gently from the higher school building down to the lower one, playfully linking the two. Grass and wild flowers grow up on the roof, and there is even a small walkway. The schoolchildren also have access to two terraces on each roof, where they can have outdoor lessons. From here there are stunning views of the town and the surrounding mountains and forests.

这三栋建筑都是木结构框架,包括电梯在内的其他所有部分也都采用了木材。学校的结构框架是靠正交胶合木立于混凝土基础之上。墙、地板和天花板则覆以预制木镶板。人们走在建筑里,所到之处都能看见裸露的木材,这种被木材环绕的感觉很美妙。人们能看到和摸到木材,也就建立了一种友好的感觉——教室和走廊散发着的温暖、坚实和安全的感觉。体育馆的巨大主体结构建立在混凝土基础上,立面内外表面都覆以当地产的道格拉斯冷杉。墙体内部是用稻草捆填充的OSB腔体。这种传统的墙体保温方式,正是建筑实践所呼吁的。

"当我们找到墙体的处理方式和比例时,我们发现这是个运用稻草的机会。这真是让人兴奋!这种天然可降解的材料也很符合项目的宗旨。" Lucas 说。

两栋学校建筑的外观——包括它们的不规则墙面、花草和V形建筑安静角落里的小空间,都非常有趣地与对称的体育馆形成了鲜明对比。虽然乍一看建筑们风格各异,但它们在视觉上却是紧密联系在一起的。不仅是因为建筑都采用了木材,在高度方面它们也有一种呼应。在场地北部,体育馆升到最高,然后通过两层楼的小学,向下延伸向西南角最低矮的幼儿园。建筑物坐落在一处坡地上,建筑师们利用这个斜坡尽量控制建筑物高度,使它们更好地与该区域的平缓低矮环境相融合。小学的正门位于山背面的底部,而学生们的储物柜则位于山坡上开凿的一块自然光线不足的区域。借助于斜坡,体育馆在整体

"We wanted the school buildings to have a strong connection with nature. And the grass-lined roofs were not meant to be seen and enjoyed just by the school's neighbours; we wanted them to be used by the pupils and teachers," explains Lucas Jollivet of architectural firm Tectoniques.

THE COMPLEX THAT TECTONIQUES DESIGNED also includes the Hacine Cherifi sports hall. When the financial crisis hit in 2009, the sports hall project had to be shelved. However, once Europe had finally recovered, the work continued and the sports hall was soon completed. The three buildings are the result of a strong belief in wood as the construction material of the future. Tectoniques has made a name for itself in France for its strict stance on environmental and sustainability issues – and for using wood wherever possible.

"We love wood. And we want it to be seen! We don't want to hide materials or make them look like something they're not. Concrete is concrete, wood is wood. The materials should establish the feel and also be practical, ideally. Having a concrete substructure for the sports hall is perfect, as balls can be kicked against the walls without damaging or marking them," says Lucas.

All three buildings are built around a frame of wood, and wood is used in practically everything else, including the lift shaft. The school's structural frame comprises cross laminated timber on concrete foundations. The walls, floors and ceilings are then lined with prefabricated wood panelling. The sense of being surrounded by wood is striking, with exposed wood everywhere you go. The material can be seen and

touched, and it creates a friendly feel – the classrooms and corridors exude warmth, solidity and security. The carcass of the sports hall rests on concrete foundations that hold up a large, solid wood structure. The façade is clad in locally sourced Douglas fir both internally and externally. Inside the walls there are OSB boxes filled with straw bales. This traditional way of insulating walls appealed to the architectural practice.

"When we worked it out and understood the scale of the walls, we saw an opportunity to try using straw bales. It felt really exciting! The biodegradable, natural material also felt very much in keeping with the ethos of the project," says Lucas.

简单的材料
体现想象力和对称

THE EXTERIORS OF the two school buildings – with their irregular façades, flowers and small outdoor spaces tucked into the quietest corner of the V-shaped buildings – playfully stand in stark contrast to the symmetrical sports hall. But although at first sight the buildings look very different, they are linked together visually. Not only are they buildings where wood has been given free rein; in terms of height they also have a kind of stepped relationship to each other. The sports hall, in the north corner of the plot, rises up highest and, via the two-storey elementary school, it all flows down to the low Preschools in the south-west. The buildings sit on a slope, something that the architects exploited to keep down their height and make them fit in better with the low-rise nature of the area. The main entrance to the elementary school building is via the lower floor at the back of the hill. And the part of the lower level that is dug into the hillside – and thus lacks natural light – is where the pupils' lockers are located. The sports hall was kept a whole six metres lower by once again using the slope. Lucas enjoyed working on every aspect of the project, but if he had to choose a favourite, it would be the sports hall.

"It may not look quite as exciting, but it typifies exactly what we want to be doing at Tectoniques. When you look at the hall, you can see with your own eyes how it's constructed. It represents a comprehensibility, clarity and simplicity that I love," says Lucas.

The finished result may look simple, but that doesn't mean it was simple to construct, at least not in the case of the sports hall roof. The enormous span required 34 metre long beams that had to be transported in two sections and put together on site. An interesting feature is the way the beams are positioned to create a kind of stepped design in the roof. This made it possible to insert large horizontal

上又低了六米。Lucas 很享受设计这个项目的每一部分，但他最喜欢的还是体育馆。

"它看起来或许不是非常吸引眼球，但它是我们 Tectoniques 事务所想做的东西。如果你仔细看这座体育馆，就能明白它是怎样被建造的。它能够被理解、清晰、简单。"Lucas 说。

最终完成的结果也许看起来很简单，但并不意味着建造过程是简单的，至少体育馆屋顶的建造不容易。大跨度需要用到 34 米长的梁，它们不得不被分成两部分运输再在现场组装到一起。一个有趣的特点就是梁搁置的方式，它打造了一种在屋顶阶梯状的设计方式。这使得在屋顶嵌入大型横窗成为可能。朝北的窗户给体育馆带来了温和持续的日光而不至于馆内人员受到太阳光的直射。学校建筑也优先保证了良好的光照和声学。大窗户不仅能完美展现户外的所有草地，而且能为教室和走廊带来日光。长长的屋檐则可以防止过强的阳光进入建筑内部使得室内过热。

windows into the roof. The windows, which face north, bring gentle, constant daylight into the hall, without the hall's users being hit by direct sunlight. The school buildings also prioritised good light – and good acoustics. As well as attractively framing all the greenery outside, the large windows also bring daylight into the classrooms and corridors. The long eaves of the roof then prevent overly strong light from entering the building and making it too hot indoors.

EVERYTHING WENT TO PLAN with the school project in Rillieux-la-Pape. The most challenging part of the process, according to Lucas, was the fact that the planning stage took such a long time.

"It takes good preparation and dialogue with the project's various craftsmen to keep everything on track. The electricians and plumbers need to explain early on where they need to run cables and pipes and so on, because you want to avoid changing too much once all the parts arrive on site. On the other hand, progress is quick once the building work gets started," states Lucas.

The schools and the sports hall were completed in 2013 and 2014 respectively, and now form a natural part of everyday life for Rillieux-la-Pape and its schoolchildren. Which brings us back to the article in the Huffington Post. The text concludes by arguing that children not only become more motivated and learn maths, language or history better in an environment where grass, trees and flowers are a natural part of the school day. They also gain a deeper understanding of nature itself. When the children get to see it and learn to appreciate and love it, a generation is created that is more inclined to look after and protect nature in the future.

Rillieux-la-Pape 镇的学校项目一切都是按计划进行的。据 Lucas 说，过程中最具挑战性的部分就是规划阶段花费很长时间的事实。

"这需要和项目的不同木匠进行充分的准备和对话才能确保一切不脱离正轨。还需要和电工和管道工尽早解释他们安置电缆和管道的地方等等，因为一旦部件都送到现场我们希望尽量避免改动太多。另一方面，一旦建筑工作启动进展就快了。" Lucas 说。

学校和体育馆分别在 2013 年和 2014 年相继完成，现在成了 Rillieux-la-Pape 镇和学校孩子们的日常生活中很自然的一部分。让我们回到赫芬顿邮报中的那篇文章。该文章在结论中争辩道，在草地、绿树与鲜花是学校生活很自然的一部分的环境中，孩子们不仅会变的更主动，更好地学习数学、语言和历史，而且对大自然本身有了更深刻的理解。这些能亲眼看到并学会欣赏和热爱大自然的孩子，正是未来更倾向于照顾和保护大自然的一代。

SCHOOL COMPLEX AND SPORTS HALL BY TECTONIQUES

Lyon-based Tectoniques has a strong focus on environmental and sustainability issues, plus specialist expertise in building with wood. Over its 20 years in the industry, the practice has gradually developed guidelines for all its projects: easily understood architecture where the materials are visible and make an impression. Unnecessarily complicated details are scaled back in favour of developable and adaptable projects.

Client: Rillieux-la-Pape Commune.
Contractor, structural fram: Favrat (schools), Lifteam (sports hall).
Cost: EUR 8.9 million (schools), EUR 3.7 million (sports hall).

Tectoniques 事务所设计的综合学校和体育馆

位于里昂的 Tectoniques 事务所密切关注环境与可持续发展问题，在木建筑方面也有很多专业经验。在行业中的 20 年来，从实践中逐渐发展出其所有项目的指南：用可视化和令人印象深刻的材料打造容易理解的建筑。不必要的复杂细节都会被缩减以利于发展性和适应性强的项目。

客户：Rillieux-la-Pape 公社
结构框架承包商：Favrat（学校），Lifteam（体育馆）
费用：890 万欧元（学校），370 万欧元（体育馆）

Pine warms futuristic landmark
松木暖热未来主义新地标

Text: Erik Bredhe | 校译：韩佳纹
摄影：Hundven Clements photography

| Church of Knarvik in Knarvik, Norway by Reiulf Ramstad
| Published 18 March 2015

The Norwegian coast has gained a new landmark. In Knarvik, the spire of the newly built wooden church rises up amongst the fjords and mountains. The doors to the church recently opened, welcoming the faithful and fans of culture into the warmth.

FOUR FJORDS MEET in Knarvik – Osterfjord running south, Sørfjord heading south-east, Salhusfjord south-west and Radfjord north-west. Only 5,000 people live in the little town far out on Norway's west coast, but it is in no way an isolated community. Two large bridges connect Knarvik with Norway's second city of Bergen. Until the last bridge was completed in 1994, a ferry ran between the city and Knarvik, carrying more vehicles each year than any other ferry route in Norway.

Since last year, there has been even more of a reason to visit the windswept little town. The newly built wooden church on the hill above Knarvik is a fascinating creation. Its pyramid-shaped spire, which echoes the surrounding mountains and fjords, became an immediate local landmark.

"The inspiration for the church came primarily from the natural features of the area. But also from the local tradition of Norwegian medieval stave churches. The church is distinctive and simple in its geometry, materials and design," says Reiulf Ramstad, chief architect at Reiulf Ramstad Arkitekter.

DESPITE ITS MODERN, almost futuristic look, the church signals its function with spiritual dignity. This architectural

mix is fully intentional, with the church built to be a place where religion and culture can come together. In recent years, several Nordic churches have similarly tried to find new ways to play a part in an increasingly secularised society. The trend for pop and rock concerts to be held in churches is becoming ever stronger, as the performers embrace the acoustics and aesthetics.

In the Church of Knarvik, the minimalist interior has a very Scandinavian feel and the use of bold Christian symbols has been reigned in. The traditional pews have been replaced with modern wooden chairs on small metal legs. The aim is to make it look and feel like a place for everyone, a space equally suited to contemplative prayer and a cultural buzz.

松木暖热未来主义新地标

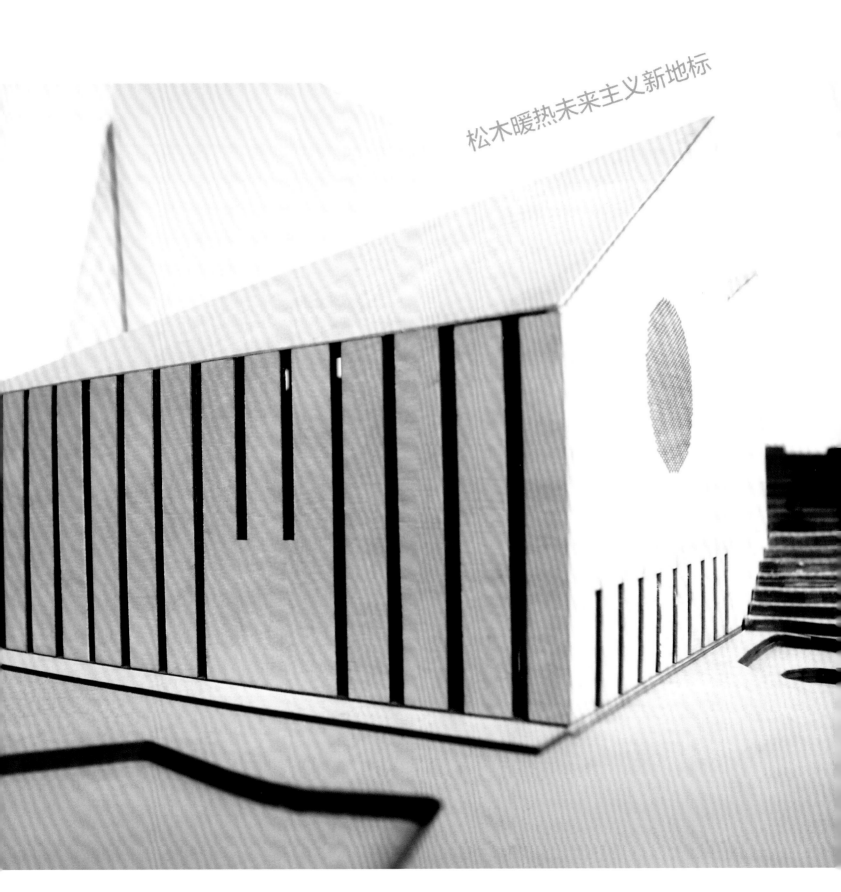

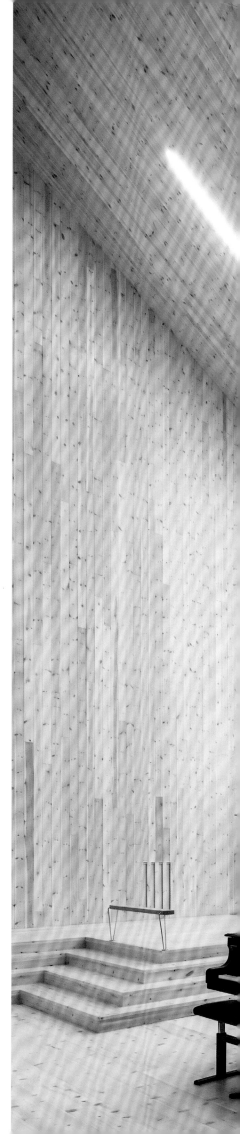

挪威海岸树立起新的地标。新建的木教堂的尖顶在 Knarvik 的峡湾和山脉之间升起。教堂已经投入使用,温暖的欢迎虔诚的教徒和文化爱好者。

四条峡湾汇聚于 Knarvi,南面的 Osterfjord、东南的 Sørfjord、西南的 Salhusfjord、西北的 Radfjord,仅有 5000 人住在这个远在挪威西海岸的小镇上,但它绝不是一个孤立的社区。两座大桥连接 Knarvik 与挪威的第二大城市卑尔根。直到最后一座大桥于 1994 年建成之前,渡轮在卑尔根市和 Knarvik 之间来回穿行,比起挪威其他的渡轮路线它每年承载更多的交通。

自去年以来,又多了一个去这座小镇的理由。Knarvik 在山上新建的木制教堂是一个令人着迷的创造。其呼应着周围的山脉和峡湾的金字塔形的尖顶,立即成为了当地的地标。

Reiulf Ramstad 参赛队伍首席建筑师 Reiulf Ramstad 说:"教堂的灵感主要来自该地区的自然特征,还有挪威当地传统的中世纪教堂。这个教堂的几何结构、材料和设计是独特的并简单的。"

尽管它是现代的,几乎有一种未来主义的倾向,教堂用一种精神性示意其功能。这种建筑上的混合是有意为之的,通过教堂建立起一个可以将宗教和文化聚在一起的场所。近年来,一些北欧教堂同样试图寻找新的方法来参与到日益非宗教化的社会中。在教堂里举行流行音乐会和摇滚音乐会越来越流行,正如演员同样学习声学和美学一样。

Pine warms futuristic landmark

"The people of Knarvik have long been looking for a way to unite the community. The new church fulfils this function and has become not only a place for religious gatherings, but also a cultural centre for art and a venue where young people can sing and learn to play instruments. The architecture of the church, the spatial solutions and the choice of materials combine religion, culture and local history," explains Reiulf Ramstad.

THE CHURCH COMPRISES two floors, with the nave on the upper floor. This is where the entrance to the church lies, accessed via broad concrete steps. Cultural activities and administration take place on the lower floor. In total, the church has capacity for upwards of 500 people. The architects were clear from an early stage that they wanted to build the church in wood.

"We chose wood mainly because it's such a traditional construction material in Norway. We know our wood here. But it was also important for the organic, living feel it conveys, and its purely aesthetic properties," states Reiulf Ramstad.

The design comprises pine cladding and pine details. Parts of the structure were prefabricated and brought to the site for assembly. A steel frame has been clad in pine panelling internally and externally. Externally, the cladding has a pre-weathered, slightly more speckled appearance, allowing it to merge superbly into its setting. Internally, a lighter option has been used. To further brighten up the church's interiors, the aim was to let in as much natural light as possible, without it causing problems if using a projector, for example. The lower level benefits from the church being on top of a hill, with the side facing towards the town getting plenty of light through its windows. During the day the small windows of the church let in just enough light and at night they provide a warm glow.

Pine warms futuristic landmark
松木暖热未来主义新地标

"We created a good balance between glass and wood that we're very pleased with. The narrow vertical windows were very carefully planned. Their shape, size and position were designed to maximise light without the risk of glare," says Reiulf.

The roof of the church comprises three flat triangles that emanate from the spire. These triangles are all gently angled upwards, towards the sky – as if opening the church up to heaven. Yet however the church was designed, here on the windy Norwegian coast, it would always have been subject to the vagaries of the heavens above. Reiulf Ramstad Arkitekter was forced to install a meticulous system to keep rain and damp away from this highly exposed wooden building.

IT TOOK AROUND two years to complete the church, in a process that was not entirely painless. Towards the end, the building contractor suffered financial problems, which brought construction to a halt until a new contractor was able to take over the reins. The economic downturn also prompted Reiulf Ramstad Arkitekter to redesign some of the details, and initially a church organ had to be borrowed from Bergen. However, spurred on by the words of former Norwegian bishop Ole D. Hagesæther that "Knarvik without a church is like a gap in the teeth, but soon we will smile in Knarvik!", the church was completed in 2014. And there is every reason to believe that not only the faithful, but also fans of architecture from across the globe, will make the pilgrimage to this unique wooden church.

在 knarvik 教堂，简约的内部非常有斯堪的纳维亚的风格，大胆的基督教符号也被运用在其中。传统的教堂长椅被现代风格的短金属腿木椅取代。其目的是从教堂的外表让每个人感觉到这是为他们提供的既适合沉思祈祷又适合文化交流的地方。

Reiulf Ramstad 解释说："Knarvik 的人们一直在寻找一种增强社区凝聚力的方式。这个新的教堂实现了这个功能，这里不仅仅是一个宗教聚集的场所，也是年轻人能够唱歌、学习乐器的艺术文化中心。建筑本身、空间解决方案和材质选择结合了宗教、文化和当地历史。"

这个教堂有两层楼，教堂正厅位于上层。通过室外混凝土大台阶，拾阶而上，就来到教堂的入口。文化活动和行政在底层。这个教堂最多可以容纳 500 人。建筑师在初期就确定了使用木材建筑建造这座教堂。

Reiulf Ramstad 说："我们选择木材主要是因为它是挪

威传统的建筑材料，我们了解我们这里的木材。它传递的有机生活的理念和纯粹的审美属性也很重要。"

这个设计包括松木的覆层和松木的细节。有部分结构是预制的，带到现场组装。钢构架的里外都由松木板包覆。松板外观已经经过预风化、带有斑点，使其能够很好地融入到环境中。在内部，设计师选择使用更轻的材质。为了进一步提亮教堂的内部，设计尽可能多地让自然光线进入，从而在使用投影机时不会产生问题。例如，教堂的底层充分得益于教堂在山顶的位置，使得面向小镇的一侧得到了充足的自然光。白天，教堂的小窗能够让足够光线进入，在夜晚，它们则带来一抹温暖的光亮。

Reiulf 说："我们高兴的是创造了玻璃和木材的合理平衡。狭窄的垂直窗户经过非常仔细的设计。形状、尺寸和位置的设计目的是达到没有避免强光的最大进光量。"

教堂的顶部是由三个从塔尖发散开来的扁平三角形组成的。这些三角形角度都轻微向上，朝着天空。就好像打开了教堂通向了天堂。但是无论教堂是怎样设计的，在风吹拂的 Norwegian 海岸，它一直从属于上帝的奇想。Reiulf Ramstad Arkitekter 被强制要求安装一个特殊设置的保护系统来让这个暴露的木结构免受雨和湿气的侵蚀。

这个教堂花费了两年的时间得以建成，在建造过程中并不是完全没有困难的。在收尾的时候，建筑承包商遇到了资金问题，这导致了建造停工了一段时间至到新的承包商接手了这个项目。资金上的滑坡也使 Reiulf Ramstad Arkitekter 重新设计了一些细节，而起初教堂的元件也必须从卑尔根借来。然而，前挪威主教 D.Hagesæther 说："没有教堂的 Knarvik 就像是一个有缺口的牙齿，不久我们都将会在 Knarvik 笑！"这个教堂于 2014 年建成。有足够的理由相信，除了虔诚的信仰者还有来自全球各地的建筑迷都将会到这独特的木质教堂朝圣。

Pine warms futuristic landmark

THE CHURCH OF KNARVIK BY REIULF RAMSTAD

The Church of Knarvik took two years to build and was completed in 2014. With capacity for 500 people in 2,250 square metres, it performs an important function as a community centre for all the town's citizens, whether they are worshippers, culture enthusiasts or people just looking for a refuge. The church was designed by Oslo-based Reiulf Ramstad Arkitekter, which has made a name for itself as a creator of innovative and daring, but at the same time simple, architecture. The practice mainly operates in Norway, but is also involved in the Region City Gothenburg project and others.
Client: Lindås Kyrkjelege Fellesråd
Cost: SEK 88 million

I'M NOT A BIG FAN OF SUBURBS AND THAT TYPE OF HALFWAY AREAS

From the spectacular visitor centre on the steep slopes of the Trollstigen road to the stylish wooden nursery in Fagerborg, Oslo – over the course of his career, Reiulf Ramstad has become renowned for his bold and simple designs with strong ties to Scandinavia.
Why did you want to be an architect?

"There are millions of paths you could take in life. Initially I wanted to be a ballet dancer, but after four years at the Norwegian National Ballet School in Oslo, I realised I was much better at architecture. I've always been keen on drawing, and I still draw every day. I think everyone should do more drawing! I think it's important, even though our work often ends up in a digital format."

What do you like best about working with wood?

"It's a way of showing respect for tradition. At the same time, wood is an innovative material with great potential for development. Wood is wonderful at conveying feelings. I mainly prefer to work with Nordic woods, but I love all types of wood and the challenge is matching the right wood to the right project."

How do you see the use of wood developing in future projects?

"Wood is a fascinating material that can be used for so much. I'd love to see more large-scale projects and buildings in wood."

What makes a good collaboration between architect, structural engineer and contractor?

"A healthy dialogue, a generous exchange of experiences and good energy. If you work with skilled, passionate people, that increases the chance of achieving a distinctive result of the highest quality."

What inspires you?

"I like urban environments and nature. I'm not a big fan of suburbs and that type of halfway house. Densely populated urban environments are exciting and give people an opportunity to live close to each other without barriers or distance. But I also like open, empty landscapes that offer time for reflection."

What projects does your practice have in the pipeline?
"We're working on 20 different projects, six of which have reached the construction phase. In Norway, we're working on a metro station in Oslo, a museum, another church and a retirement home. In Sweden, we're one of several architectural practices working on the Region City project in Gothenburg. The focus is on developing the area around the Central Station and Älvstaden. Swedish and Norwegian culture has a great deal in common and we really enjoy working in Sweden."

（KNARVIK 教堂——建筑师 REIULF RAMSTAD）

Knarvik 教堂建造了两年，在 2014 年建成。能够在 2250 平方米的地方容纳 500 人，它的一个重要的功能是作为所有市民的交流中心，无论市民是宗教信仰者、文化爱好者，或者仅仅是来寻找一个避难处的人。奥斯陆建筑师 Reiulf Ramstad 设计了这座教堂，一座创新和意气风发的建筑。他的主要工程在挪威，但也涉及了哥德堡项目及其他地区项目。

客　户：Lindås Kyrkjelege Fellesråd
总造价：SEK 88 million

我不是一个喜欢郊区和那一类中间状态地区的人
从 trollstigen 道路陡峭山坡上壮观的游客中心到 fagerborg 的风格化的木制幼儿园，在 Reiulf Ramstad 的整个职业生涯中，他因为与斯堪的纳维亚关系密切的大胆和简单的设计而闻名。

为什么你想成为一名建筑师？
在一生中有很多路可以选择。起初，我想要成为一名优秀的舞者，但是四年之后在奥斯陆的 Norwegian National Ballet 学校，我意识到我更擅长建筑。我一直渴望画画，我每天都在坚持画画。我认为每个人都应该画很多画！我认为画画是重要的，即使我们的工作经常以精确的数据收尾。

你最喜欢的是使用木材中的那一部分？
这是一种尊重传统的方式。同时，木材是一种具有巨大发展潜力的创新材料。木材可以很好的传达情感。在工作中我更喜欢北欧的木材，但是我喜欢各种类型的木材，和迎接合适的木材匹配在合适的项目的挑战。"

你怎样看在未来项目中使用木材的发展前景?

木材会被大量的使用,因为它是吸引人的材质。我很乐意看到越来越多的大型工程和建筑使用木材。

建筑师、结构工程师和承包商之间怎样建立好的合作?

"友好的对话,慷慨的交流经验和充沛的能量。如果你与技能熟练的、充满激情的人一起工作,将会增加得到最高品质结果的机会。"

是什么给了你灵感?

我喜欢城市的环境和生态。我不喜欢郊区和那种类型小房子。人口稠密的都市环境是令人振奋的,它提供给人们没有障碍和距离的深入接触的机会。但是我也喜欢开阔、空荡的空间,这样的空间带给我时间来反思。

Pine warms futuristic landmark

现在你有哪些项目正在进行？

"我们正在做 20 个不同的项目，其中的六个已经到了建造阶段。在挪威，我们正在奥斯陆建造地铁站，一个博物馆，另一个教堂和一个养老院。在瑞典，我们是参与哥德堡地区城市项目建设的建筑公司之一。重点是发展中心站周围的地区和 Älvstaden。瑞典和挪威文化有很多共通之处，我们真的喜欢在瑞典工作。"

Japanese harmony between building and nature
建筑与自然之间日式和谐

Text: Erik Bredhe | 校译：高瑜
摄影：Koji Fujii/Nacasa&Partners;Hiroshi Nakamura

| 狭山教堂；日本狭山；中村拓志
| 2015年12月7日出版
| Sayama chapel in Sayama, Japan by Hiroshi Nakamura
| Pubished 7 December 2015

The little chapel next to the cemetery in Sayama looks like it has risen out of the ground. The larch structure creates a space for contemplation and a tranquil interaction with nature.

WITH HIS TSUNAMI-PROOF coastal evacuation tower, tepee house and mystical treehouses, Hiroshi Nakamura has established himself as one of the most distinctive and inspiring voices in Japanese architecture. With the new chapel next to Sayama cemetery, he continues to forge his own path. The little chapel shoots up in a leafy area of forest like something out of Hayao Miyazaki's beautiful animated films.

"I wanted the chapel to be a place for people of all faiths. So I thought about what all religions have in common and what came to mind was prayer. The action of the two hands became the starting point for the chapel," says Hiroshi Nakamura.

In Japan, the word 'gassho' describes both two hands placed together in prayer and a type of architecture. Farmhouses in gassho style are a common sight in the Japanese countryside. They are a hangover from the Edo period, when many farmers built the pointed roofs so they could grow silkworms in the attic.

THE STRUCTURE COMPRISES 251 pairs of beams made from Japanese larch that lean in towards each other in an inverted V. The long beams, some as much as nine metres, were fixed together with metal ties at the top before they were erected. When the chapel was assembled, they were first lifted up and then lowered down into the right position. The beams were then secured to the ground by running metal pins sticking up from the large base plates up into drilled holes in the endwood.

"This method required incredible precision, not least because the structure is so exposed, without any design details to conceal mistakes. But we created everything in a 3D environment and in the end had a tolerance of just one millimetre."

BY DEVELOPING THE CHAPEL'S gassho style and having the building project outwards in several different directions, the structure is able to handle both horizontal and vertical stresses – giving protection against earthquakes. The roof is also clad in four millimetre thick sand-cast aluminium plates. This was the thinnest size possible for sufficient durability, and the thickest possible so that the plates could still be bent by hand to make them fit the curved roof.

建筑与自然之间日式和谐

Japanese harmony between building and nature

这座小教堂在峡山墓地旁破土而出。这个落叶松木建筑打造了一个可以沉思、可以与自然交流的空间。

抗海啸沿海疏散楼、美式圆锥形帐篷住宅和神秘的树屋,中村拓志的作品已然让他本人具有日本建筑界最与众不同、最具影响力的发言权。在峡山墓地旁新造的教堂继续开拓他的设计风格。这座小教堂处在一片绿树掩映的森林之中,仿佛来自于宫崎骏唯美的动画电影。

"我希望这个教堂可以为各种信仰的人敞开大门。我仔细思考以后发现,所有宗教的共同点和所思所想都是祈祷。双手合十的动作即为教堂灵感的启发点," 中村拓志说。

在日本,"gassho"这个词既描述了双手合十祈祷的动作,也是一种建筑类型。gassho 风格的农舍在日本乡下非常常见。这些农舍是江户时期的遗存,那时候农民们建造尖顶屋子以便在阁楼上养蚕。

这座小教堂由 251 对日本落叶松木梁构成,梁与梁承倒"V"字形相互向内倾斜交错。这些梁有的高达 9 米,架起来之前,需要用金属条连接顶端节点。在搭建教堂时,首先将梁抬高,再扎入合适的位置。然后用金属栓从大底盘往上打入木梁底端的钻孔,将梁牢固在地上。

"此方法需要极高的精度,尤其是因为结构一览无余,没有任何设计环节可以隐藏失误。我们的设计用全 3d 进行,最终仅容有 1 毫米的误差。"

建筑与自然之间
日式和谐

Inside, the floor is laid with natural stone. Nakamura chose to keep the chisel marks from the shaping of the flagstones. As with the aluminium roof in its unpolished state, they were chosen to indicate some kind of relationship between humanity and the forces of nature.

"We worked actively to bring sunlight and other natural phenomena into the interior of the chapel through the large triangular windows. But also by expressing the proximity to the forces of nature via the structure of the materials. I thought this felt appropriate for a chapel."

THE FLOOR SLOPES GENTLY towards the forest. The scratched structure of the flagstones also guides the eye towards an imaginary point deep in the dense forest. The chapel merges in with the beautiful, tranquil surroundings, forming a place where the grief and pain of visitors is handled with great sensitivity.

"The chapel is not meant to be an architectural showpiece. It's a consequence of its surroundings. A place close to nature is a place close to the soul," concludes Nakamura.
HIROSHI NAKAMURA

The connection between a building and nature is important to Hiroshi Nakamura, as can be seen in the chapel at Sayama cemetery. He allowed the location to interact with the chapel through light, materials and natural finishes. To strengthen the connection between the building and nature, he first planted new trees on the triangular plot – and then created the chapel around them. The inward slope of the chapel's roof allows space for the branches of the trees to grow.

Japanese harmony between building and nature

小教堂通过"gassho"结构形式以及朝多个方向的外向延展,足以承受水平和垂直两个方向的应力——足以抵御地震。屋顶上也镀了一层四毫米厚的砂铸铝板。此为经久耐用的最小值,也是人工可弯曲的最大值。

建筑内部的地面铺设了天然的石材。中村拓志保留开凿石板时的凿痕。和未经打磨的铝材屋顶一样,它们的设置象征了人类与自然力量之间的关联。

"我们极力将阳光和其他自然现象通过三角形的大窗户透进教堂。同时也通过各种材质的结构表达对自然力量的亲近。我觉得这很适合一个教堂。"

教堂地板慢慢向森林倾斜。石板上的凿痕结构也将视线逐渐引向森林深处。教堂与美丽、宁静的环境融为一体,让所有祭拜者的悲伤和痛苦都随之萦绕开去。

中村拓志最后说:"这座教堂并不是一个观赏性建筑,它与周边环境浑然一体。亲近自然之地即靠近灵魂之所。"

HIROSHI NAKAMURA

The connection between a building and nature is important to Hiroshi Nakamura, as can be seen in the chapel at Sayama cemetery. He allowed the location to interact with the chapel through light, materials and natural finishes. To strengthen the connection between the building and nature, he first planted new trees on the triangular plot – and then created the chapel around them. The inward slope of the chapel's roof allows space for the branches of the trees to grow.

中村拓志

对中村拓志来说,建筑与自然之间的联系是很重要的,这一点从峡山墓地的教堂可见一斑。他允许教堂的所在地通过光线、材料和自然的修饰来与教堂进行互动。为了加强建筑与自然之间的联系,他首先在三角地块上种植新树,然后在树的周边建造了教堂,教堂内斜坡的屋顶有足够的空间留给树枝生长。

Architects Promote Books
建筑师是书的推手

Text: Leo Gullbring ｜ 校译：蒋音成
摄影：Rasmus Norlander

| Book Mountain in Spijkenisse, the Netherlands by MVRDV
| Published 22 November 2012
|书山图书馆，斯皮让塞，荷兰
|建筑设计：MVRDV

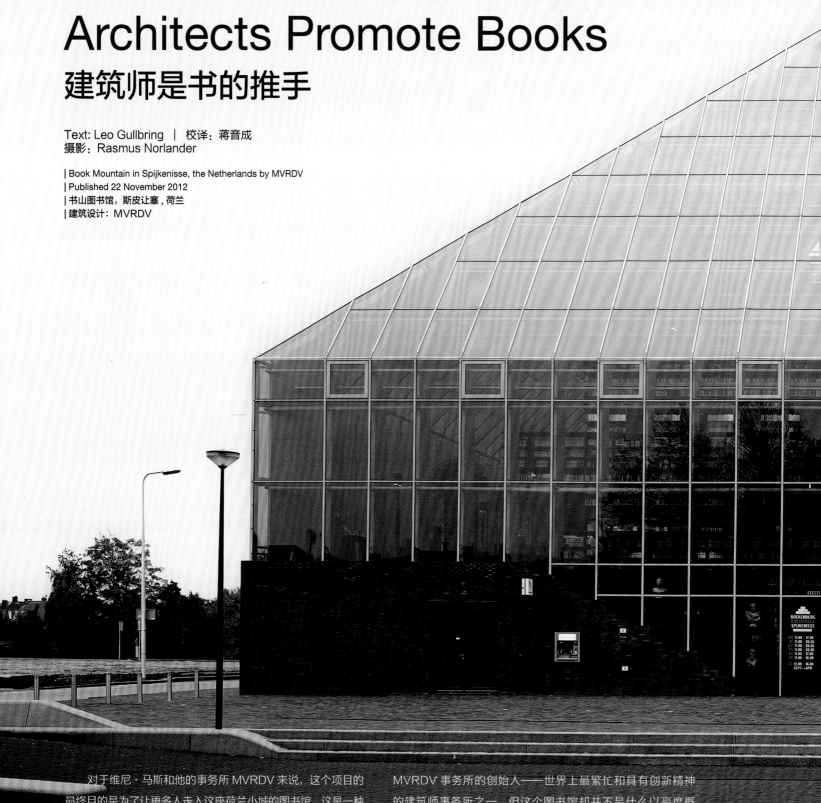

对于维尼·马斯和他的事务所MVRDV来说，这个项目的最终目的是为了让更多人走入这座荷兰小城的图书馆，这是一种较为戏剧化的方式，即在这座海港城市的中心把书堆积起来。"我们选择在一个完整的大空间展现这些书籍宏大而美好的一面，我们的设计是一座可以攀爬的书的山，每个人都可以可看并且可以爬山。我们的图书馆一种对于书的纪念，在这里，书作为一种文化表达的方式，当然也是一个让人聚集和存储档案的地方。"

Winy Maas（维尼·马斯）有如此热情不足为奇，他是MVRDV事务所的创始人——世界上最繁忙和具有创新精神的建筑师事务所之一。但这个图书馆却并不是什么以高度概念化讲求排场的项目，相反，这是一个接地气的项目：据说这座小城仍有百分之十的居民是文盲，而图书馆目的就是让这里人们多来这里看书。"我们将建筑的室内空间当做室外空间来做，因为我们希望它成为小城的中心，而不是一个普通而中立的建筑，通过这个建筑我们试图提升书在社会中的角色。我们保留了砖砌的老教堂结构作为基座，被木材和玻璃包覆。"

Kommer boken att överleva eller försvinna? Arkitekten Winy Maas från MVRDV ställer frågan och presenterar biblioteket Bookmountain i Spijkenisse. Även kinesiska och norska arkitekter hyllar den traditionella bokkonsten med hjälp av uppseendeväckande träarkitektur.

WINY MAAS AND his architectural firm MVRDV wanted to increase visitor numbers at the library in the Dutch town of Spijkenisse, and chose to do so in a dramatic manner – by piling up a stack of books in the middle of the port town.

"We chose to display the books in their full glory in one large space, a whole mountain of books to climb, which everyone can see and access. Our library is a monument to the book as a means of cultural expression, as well as being a meeting place and an archive."

There is no faulting Winy Maas' enthusiasm. He is the founder of MVRDV, one of the world's busiest and most innovative firms of architects, but this was not some high concept prestige project. It was about something as down to earth as getting people to read in a town where 10 percent of the population are thought to be illiterate.

斯皮让塞以前是鹿特丹远郊的一个渔村。20世纪50年代，随着鹿特丹这一欧洲第一大港的扩建，渔村逐渐演变成了有着8万5千居民的小镇。维尼的同事Jan Knikker介绍说，从12年前起，鹿特丹宜居党开始治理鹿特丹，他们努力地进行了一些城市复兴项目，从文化和商业上重新激活这座工人阶级的小城。"他们首先建了一个后现代迪士尼风格的露天市场，现在又有了图书馆和新住宅区，再过几年Ben van Berkel新剧院又会完工。"这座新开放的图书馆坐落于老城中心，靠近主教堂，从集市广场可以清晰地看到。与新建的住宅区一起，中心城区具有清晰的实体感，图书馆似乎是通过内外翻转从而暴露内部。项目主创建筑师福克尔说，在图书馆里的楼梯体验很像迷宫，人们通过楼梯爬上爬下，到达不同的楼层。"书山"被格栅状的胶合木梁结构所围绕，爬上顶更可以一览斯皮让塞的城市风景。

Architects Promote Books

"当这些巨大的胶合木梁搭好时，你应该已经能够看到图书馆的模样了。"福克尔略为惋惜地说，"立面采用玻璃几乎可以说是一种遗憾。将书籍暴露于公众视野能吸引读者来到图书馆，如果不出意外，这里还将成为老城中心一个人们聚集交流的场所。"馆内的书架高达数米，但所有书籍的借阅都十分方便可达。其他如信息服务台等设施，由旧花盆排列隔开，形成了员工区域和阅读区域。图书馆内还有一个提供报刊杂志的咖啡厅和一个儿童区。图书馆的下面是商店、办公和当地象棋俱乐部。斯皮让塞图书馆竭尽所能地提升人的文化素养并鼓励阅读。他们最近收获了一个国际营销奖，由于他们发放出一些卡片给那些没有注册他们图书馆的居民，卡片上写着，"我们想念你！"。虽然这是一个令人惊叹的建筑，但是图书馆是非常传统的，维尼强调，虽然数字技术和平板电脑的快速发展，但是他们选择去展示很多经典的印刷书籍。这种传统的方式同样在这个由从谷仓汲取灵感的建筑中有所表现。有48座MVRDV设计的建筑，都是从农业谷仓汲取灵感，图书馆是其中一座。"谷仓"现在已经成了荷兰当代建筑中一个流行的主题。

"如果想将阳光引入建筑一层，谷仓是个很好的切入点。你可以把街道规划得相对狭窄些，营造亲切和紧密的城市氛围。"维尼说，"在斯皮让塞，中心区域相当于一块农场大小，甚至还遗留了几个谷仓。设计图书馆时，我们采用了荷兰传统谷仓的形状，并进行了一个当代性转化。玻璃的立面就是一层轻薄的外壳，谦逊地折射出建筑的过去。"

147

"We're turning the library inside out. We're making it a main attraction in the heart of the town! Instead of creating a neutral building, we've chosen to promote the whole issue of the book's role in society. The brick architecture of the old church becomes a pedestal for a book mountain clad in wood and glass."

THE FORMER FISHING VILLAGE of Spijkenisse is located on the outskirts of Rotterdam. With the expansion of Europe's largest port, the village has been transformed since the 1950s into a town with 85,000 inhabitants. Winy's colleague Jan Knikker explains that the party of discontent which took power 12 years ago, the Liveable Rotterdam party, has worked hard to bring about several urban renewal projects to re-energise the working class town, culturally and commercially.

"Firstly, an outdoor market was created in post-modern historical Disney style. Now we've opened our library together with new housing, and in a few years Ben van Berkel's new theatre will also be added."

The recently opened library is situated in the old town centre, close to the main church, and stands clearly visible from the market square. Together with the newly built housing, the centre has gained a clear materiality, with the library looking like it has been turned inside out to expose its contents. According to project architect Fokke Moerel, there is quite a labyrinth of stairs running up and down, from floor to floor. Book Mountain is encased in a trellis-like framework of glulam beams, and at the very top there are views of the Spijkenisse townscape.

"You should've seen what it looked like when the large glulam beams had just been installed," says Fokke wistfully. "It was almost a shame to erect the glass facade. Putting the books on public view attracts readers to the library, and if nothing else it will become a new meeting place in the old centre of Spijkenisse."

The shelves reach heights of several metres, but all the books for borrowing are easily accessible. Beyond the bookshelves which, like the information desks and other items, are made from old flower pots, lie the staff areas and reading desks. There is also a café here with newspapers and magazines, and a children's section. Below the building itself, there is space for shops, offices and the local chess club.

SPIJKENISSE'S LIBRARY DOES everything it can to raise literacy and encourage reading. They recently received an international marketing award for having sent out postcards carrying the text 'We miss you!' to residents who were not on their borrower's register. Despite its spectacular architecture, the library is quite traditional, and Winy Maas stresses that they have chosen to showcase the classic printed book, despite the proliferation of tablets and other digital technology. The traditional approach is also reflected in the barn-inspired architecture. As well as the library, MVRDV's 42 newly built homes reflect the old agricultural barns, something that has become a popular theme in contemporary Dutch architecture.

建筑师是书的推手

"The barn is a great starting point if you want to create homes with lots of light at ground level, and then you can develop the town plan with narrow streets and an intimate, concentrated urban character," says Winy. "In Spijkenisse, the central district is about the size of a farmyard, and there are even a few barns left. We've taken the shape of the Dutch barn and given it a contemporary twist. The glass facade of the library acts as an incredibly thin shell reflecting the architecture of the past, but without being too obvious."

Winy emphasises that he is not advocating some sort of neo-traditionalist, neo-urbanist nostalgia for the past. However, MVRDV does want to draw on an older architectural vernacular to create future solutions. The idea behind Book Mountain was to scale up the barn and generate curiosity about its content. Winy cites the story of Alice in Wonderland as one of his inspirations.

Building in wood is as much a technical as it is an architectural choice. Winy explains that the lattice of glulam beams does more than reduce the direct sunlight. The deep beams also frame the view, giving a new expression to the town centre. In recent years, wood has gained in popularity among leading Dutch architects.

"We eliminate almost four fifths of the direct sunlight, but in addition wood absorbs heat better than steel and so acts to even out the indoor climate. Wood also makes the library environment feel less like a factory and more like a home."

Fokke Moerel adds that the wood engineers managed to replace the designers' steel plates with wooden joints, something that further accentuates the barnlike proportions. The library measures almost 10,000 m² and has an extremely efficient and innovative climate control system.

"Construction work on Book Mountain began before BREEAM certification came in," states Fokke Moerel, but explains that the building was created with the least possible environmental impact and comfortably meets all environmental requirements.

"Instead of air conditioning, we have natural ventilation. We use rainwater run-off from the roof for toilets and sprinklers, and we also have PCM gel in the offices that, like wood, evens out temperature peaks and troughs. Under the building there are also large wells that act as energy reserves for both heating and cooling."

Spijkenisse's Book Mountain was inaugurated on 4 October 2012 and has already been criticised for a kind of book fetishism. But why not? A mountain of books to explore and borrow is perhaps just what is needed for a renaissance in reading.

维尼强调，他并不是在推崇新传统主义、新都市主义的乡愁情结。但 MVRDV 确实是希望从更有历史的乡土建筑中探索未来建筑的答案。书山图书馆的设计想法是增加谷仓的尺度。维尼还提到《爱丽丝梦游仙境》是他的灵感来源之一。

采用现代木结构既是建筑设计的需要，也是技术上的选择。维尼解释道，格栅状的胶合木梁不仅减少了太阳光直射，还起到了框景的作用，是对市中心的一种新的表达。近年来，现代木结构颇受荷兰顶级建筑师们的青睐。"我们消除了约 4/5 的太阳光直射，此外木材比钢材能更好地吸收太阳热量，并调节室内微气候。木材也将图书馆的室内环境营造得如家庭般温暖，而不像是冷冰冰的工厂。"福克尔补充说道，木结构工程师用木结构连接件取代了设计师的钢板设计，更加强调谷仓的比例。这座图书馆建筑面积约 10000m²，室内气候控制系统相当高效和先进。"施工作业在 BREEAM 认证（英国建筑研究院环境评估）引进之前就开始了。" 福克尔说道。但这座建筑以创造对环境影响最低的建筑为目标，最终满足所有环境评估要求的结果。"我们采用自然通风取代空调系统；我们收集屋顶雨水服务于卫生间和灭火装置；我们在办公室运用相变凝胶（PCM gel），它能像木材一样平衡温度的峰谷波动。在建筑底下还有一些大的蓄水井作为能源储备供采暖和制冷。"

Spijkenisse's 与 2012 年 10 月 4 日正式开放，曾被批评为对书籍的盲目崇拜。但为什么不呢？书山有路勤为径，砌一座真正的书山供人们探索和借阅，也许正是阅读复兴所需要的。

Architects Promote Books

建筑师是书的推手

MVRDV

Dutch architects MVRDV are among the leading lights of the international architectural scene, led by founders Winy Maas, Jacob van Rijs and Nathalie de Vries.
They like to test the limits of what is possible, for example in the visual art/architectural study Pig City, where they designed 80 metre-high skyscrapers for organic pig farming. In Valencia, they are building a vertical suburb, Torre Huerta, a high-rise tower of apartments with gardens. In 2000, MVRDV also came up with a proposal for a new head office for Posten in Solna, Sweden.
MVRDV was founded in 1991 and is based in Rotterdam. The name MVRDV is made up of the initials in the architects' surnames. Today, the firm is considered one of the more influential names in contemporary architecture.

MVRDV

荷兰 MVRDV 建筑师事务所是国际顶尖的建筑师事务所之一，由创始人 Winy Maas，Jacob van Rijs 和 Nathalie de Vries 领衔。

他们喜欢尝试可能的极限，例如在"猪之城"视觉艺术和建筑研究中，他们设计了 80 米高的有机猪农场摩天楼。在巴伦西亚，他们正在建设一个空中郊区建筑，叫 Torre Huerta，一座带空中花园的高层公寓。2000 年，MVRDV 又为瑞典索尔纳邮政设计了一座新总部办公楼。

MVRDV 成立于 1991 年，总部设在鹿特丹。MVRDV 是三位创始建筑师的姓的开头字母缩写。如今，他们已是当代建筑界最有影响力的事务所之一。

世博 2015——米兰之木
Expo 2015 – Wood in Milan

Text：Leo Gullbring | 校译：韩佳纹
摄影：Sergio Grazia, Atsushi Kitagawara, Tom Vack, Marco Jetti, marcojetti.com, Andrea Bosio

Expo 2015 in Milan, Italy by Olika
Published 24 September 2015
2015 意大利米兰世博会
2015.9.24 发表

世博 2015——米兰之木

The theme for the Milan Expo 2015 is as tropical as it gets: energy and food. Wood has proven an obvious choice in many of the 54 pavilions.

Text Leo Gullbring

HOW ABOUT A 1,6 km long conveyor-belt pizza with 2 tonnes of mozzarella and 700 kilos of tomatoes? Expo 2015 is a culinary competition, where 145 participating countries, companies and organisations try to outdo each other. The two themes of the World Exhibition 'Feeding the Planet' and 'Energy for Life' are, however, more serious, what with the threat of climatic disaster and overpopulation.

It is hardly surprising that wood features in countless pavilions. After all, nature's building material goes hand in hand with a sustainability approach and gastronomic culture. China engaged Tsinghua University, and former student Yichen Lu of US firm Studio Link-Arc, to bring traditional Chinese architecture up to date with a billowing lightweight roof in bamboo – a cloudlike structure that shows how city and country can live in harmony.

Japan's Atsushi Kitagawara was inspired by traditional Kyoto houses for his highly earthquake-proof design with no metal parts, and a wall of sake barrels in the entrance. Spanish architectural practice B720 Fermin Vazquez Arquitectos has served up a prefabricated pavilion made from glulam beams which, according to the architects, has considerably reduced transport needs and created a flexible, easily assembled structure. The enormous greenhouse has energy-efficient natural ventilation and polycarbonate instead of glass.

THE CHILEAN PAVILION, just next to the concert arena, by Undurraga Deves Arquitectos goes even further, and looks like a large wooden lattice box raised on six steel supports. According to lead architect Sebastián Mallea, wood was a natural choice for a transparent experience from both inside and out, providing different types of scale with which to relate.

"As well as having the lowest possible environmental impact, wood is also a generous material that allows a flexible room design and is suitable for load-bearing, dividing and as a surface finish. It works superbly in prefabricated systems, and is easy to put together and take apart, he continues.

"We chose Monterey pine from controlled sources in southern Chile, because it has a warm quality and gives a very particular light. The smell also helps with the overall appreciation of the environment. We've used as few metal ties as possible, and most of them have been embedded for aesthetic and fire safety reasons."
If the Chilean pavilion is rather conventional, Italian architect Michele de Lucchi has pulled out all the stops for the Expo's entrance building, and above all the

Expo 2015 – Wood in Milan

2015年米兰世博会的主题是一个热门的话题：能源与食物。在54个展馆中，木材成了一种自然而然的选择。

试想一下一个在1.6千米长传送带上的披萨，用了2吨马苏里拉奶酪和700公斤番茄如何？

2015年世博会就像是一个厨艺竞赛。参加的145个国家、公司和组织都尝试着要超过对方。而世博的两个主题："滋养地球"和"生命的能源"，在气候灾害带来的恐慌和人口过载的事实下，则显得更加严峻。

众多展馆以食物为特色，这一点都不令人惊讶。毕竟，天然的建材与可持续性道路和美食文化是携手并行的。毕业于清华大学并于美国建立Studio Link-Arc的陆轶辰设计了馆。该作品将传统中式建筑通过竹制的轻型顶的演进展现了与时俱进的生命力——如同巨大波浪一样的云雾状结构，展现了城市与乡村的和谐共生。

日本的北川原温受到传统京都民居的启发，做出了没有金属构件的具有高抗震性的设计，并且在房屋入口处设置了一面用清酒桶做成的墙。西班牙的建筑事务所B720 Fermin Vazquez Arquitectos创造了附加胶合木梁预制的"亭子"，这种建筑被公认为减少了运输需求，并且创造了一种灵活易组装的建筑体系。这个巨大的温室拥有高效率的自然通风设备，并且用聚碳酸酯代替了玻璃。

Undurraga Deves Arquitectos所设计的智利馆，紧邻音乐会的场地。该建筑在这方面做的更深入，看起来像是一个巨大的由钢结构支承的木制格框。从主设计师Sebastián Mallea处了解到，如果要达到由内而外的通透体验性，木材是

strategically placed Zero pavilion, measuring 8,000 square metres, whose name reflects the UN's Zero Hunger Challenge.

"We almost always work in wood," explains project architect Angelo Micheli. "The spruce was harvested in Trentino, where wooden architecture is much more taken for granted than in the rest of Italy. Zero takes the form of a number of 20 to 26-metre tall hills, which are meant to show humanity's relationship with nature since time immemorial."

The Zero pavilion is clad in spruce boards that are supported by a metal structure. Internally, it is like a cut-out of the Earth's crust, with dozens of caves inspired by the Eugenia Hills between Padova and Vicenza. Wood also figures in a Noah's Ark and a 23-metre tall Tree of Knowledge that breaks through the ceiling.

"The structure has been driven here directly from the sawmill and assembled in the simplest way, reinforced with a steel frame. The lanterns serve as heat vents and lower the indoor temperature four degrees without mechanical ventilation."

THE MOST ADVANCED offering is the French pavilion, which Agence XTU's architects Anouk Legendre and Nicolas Desmazières have designed as a giant pergola for climbing plants and vegetables. The boxlike ceiling

一种非常自然的选择,它提供了不同的规格,而不同规格的木材可以相互连接。

"木材不仅对环境的影响产生最低的影响,也是一种优异的材料——既可以用作灵活的房间设计,也适合于承重、分隔,并且具有完成度较高的表面。它在预制系统中的发挥非常出色,并且非常易于组装和拆卸。"他继续说到。

"我们选用了来自智利南部受到控制的蒙特雷松木,因为它质地温暖,并且提供一种特别的视觉体验。它的气味增强了对自然环境的整体感受。我们尝试了尽可能少地使用金属连接件,它们的置入大多是为了美观和防火要求。"

如果说智利馆是非常传统的,意大利建筑师 Michele de Lucchi 巧妙地将零号馆放置于场地中,占地 8000 平方米。它的名字反映了联合国的零饥饿挑战。

"我们几乎总是与木材打交道，"项目建筑师 Angelo Micheli 解释道。"云杉是特伦托产的。相较于意大利的其他地区，在那里，木质建筑被视为理所当然的存在。零号馆的形象采用了 20-26 米高的山丘，旨在展现从史前开始的人与自然的关系。"

零号馆由云杉板包覆，并由金属结构支撑。在内部，它像是地壳的一个剖面，在里面有许多洞窟。这个灵感来自位于帕多瓦和维琴察之间的丁香山。木材同样出现于诺亚方舟的故事和一个 23 米高的破坏了天花板的知识树的故事。

"这个构件是由锯木厂直接运送过来的，并且用一种最简单的方式组装，用一个金属框进行加固。灯笼作为热通风口使用，并且在没有机械通风的情况下降低了室内温度约 4 摄氏度。"

最先进的建筑是法国馆。它是由 XTU 的建筑师 Anouk Legendre 和 Nicolas Desmazières 设计的一座巨大藤架，可供攀缘植物和蔬菜生长。像盒子一样的天花板被象征这个国家美食文化的绝妙样品填满。

"木材比其他材料更具有可持续性，而且很明显是一个非常符合今年主题的选择。"Anouk 和 Nicolas 给出了相同的解释。"我们的胶合木梁由质量很轻的柱形构件支撑着，这种完全由法国产木材制成的构件的内部为云杉，外部为落叶松。"

这个弯曲的设计是源于对 Pier-Luigi Nervi 和 Félix Candela 的回想，但是在这里却是由精确切割的木材构件，以恰到好处的角度相互紧扣在一起组成的。将对钢连接件的需求减到最小。因此提出了一个想法，即在世博会结束之后，将这座展馆拆卸下来，并让其在法国焕发新生。

Expo 2015 – Wood in Milan

has been filled with wonderful examples of the nation's gourmet culture.

"Wood is more sustainable than other materials and the obvious choice to fit in with this year's theme," explain Anouk and Nicolas in unison. "Our glulam beams supported by columns create a lightweight structure made entirely from French wood: spruce for the interior and larch for the exterior."

The curving and embracing design is reminiscent of Pier-Luigi Nervi and Félix Candela, but here it is made entirely in precision cut wooden components that lock each other into place at right-angles, minimising the need to add and embed steel ties. The idea is that after the Expo, the pavilion will be taken down and given a new life in France.

"We achieved this almost plastic structure with the help of industrial robots."

EXPO 2015 IS NOT without its critics. Architect Stefano Boeri and his colleague Jacques Herzog produced the Roman-inspired design for the site: a 1.5 kilometre long decumano (main street) and a number of cardine (cross-streets) housing the pavilions.

"We wanted to do something different from a typical 20th-century World Exhibition, with all the pavilions in rows," explains Stefano. "Every country would have a plot of land to showcase their best agricultural

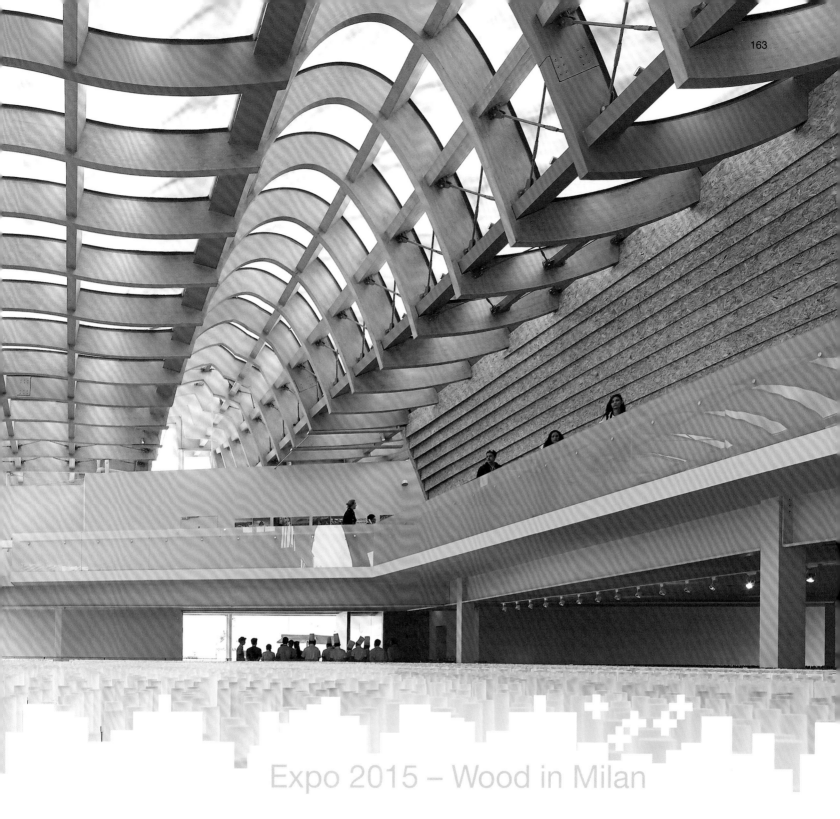

Expo 2015 – Wood in Milan

"我们在工业化机器人的帮助下完成了这作可塑性极强的建筑。"

2015年的世博会并不是没有批判的声音。建筑师Stefano Boeri和他的同事Jacques Herzog以罗马为灵感来源了设计了这样的场地——1.5千米长的主街道和若干交叉路连接各个展馆。

"我们想做出一些与典型的成排建成的二十世纪展览馆不同的东西出来。" Stefano 说到,"每个国家都应该有一块经过规划的土地去展示他们最好的农产品和先进的生物气候解决方法,以巨大温室的形式。这次世博会聚焦在我们如何在这个星球上生存下来,结果我们做成了一个大型食品市场。"

Stefano 差一点就在2011年成为米兰市长了。他是Bosco Verticale 的支持者,即设计了一些"垂直森林"居住高塔。他的瑞士同事Jacques Herzog是在国际建筑行业享有更大名气的人。当这个团队失去了对这项工作的控制时,Jacques 面向媒体表达了不满。

products, and advanced bioclimate solutions in the form of large greenhouses. An Expo with a focus on how we can survive on this planet. What we ended up with was basically a big food market..."

Stefano almost became mayor of Milan in 2011 and is behind Bosco Verticale, a pair of residential towers that create a 'vertical forest'. His Swiss colleague Jacques Herzog is an even bigger name on the international architectural scene. When the team lost control of the job, Jacques vented his fury in the press.

"We accepted the invitation to design Expo 2015 on condition that we could give it new content, not more of these monuments to national pride that have defined world exhibitions since the mid-20th century. Now Expo 2015 is just another vanity project filled with spectacular buildings."

世博 2015——米兰之木

"我们接受2015世博会的设计邀请,是建立在我们可以给予它新的内容的基础上,而不是去增加这些象征着国家荣誉的纪念碑,那是从二十世纪中期开始就对世界博览会下的定义。如今2015世博会只是另一个充斥着壮观建筑的虚荣项目而已。"

然而,慢食邀请到了Herzog&de Meuron去设计他们的展馆。这能否重新建立起他们理想中的精神?三座展馆中由敦实的柔角的胶合木梁为所有慢食的地方食物提供了一个简朴节约的背景。轻快与简洁就像是伦巴第的老旧谷仓,像是一处乡村的食品市场。这个展馆展示了濒临绝迹的食物并且提升了感官上的快乐。

"我们选择了木材是因为其具有可持续性、易加工性以及它们很适合于预制生产,"由Liliana Amorim Rocha领导的Herzog & de Meuron's的设计师团队这么解释到。"一方面是因为我们只有两个月去完成这个建筑,但因此我们也可以在秋天结束的时候将其拆卸。我们选择胶合木是因为它们极其的坚韧和牢固,具有非常出色的结构性能。"

世博 2015——米兰之木

SLOW FOOD DID, however, invite Herzog & de Meuron to design their pavilion. Would it be possible to recreate the spirit they had dreamed of? Three pavilions built from chunky, gently angled glulam beams provide a spartan backdrop to all Slow Food's sensational regional foods. Airy and simple like the old barns of Lombardy, like a food market out in the countryside. The pavilion promotes endangered foods and sensual pleasure.

"We chose wood because it's sustainable, easy to work with and suitable for prefabrication," says Herzog & de Meuron's team of architects, led by Liliana Amorim Rocha. "Partly because we only had two months to complete the build, but also so we could disassemble the pavilion at the end of the autumn. We chose glulam because it's extremely tough and stable, with excellent structural properties."

钻石声效
Diamond acoustics

Text: Erik Bredhe | 校译：韩佳纹　朱志军
摄影：Marie-Caroline Lucat

| 柏林犹太博物馆，德国，丹尼尔－李伯斯金
| 2014年3月14日出版
| Théâtre Jean-Claude Carrière in Montpellier, France by A+ Architecture
| Published 14 March 2014

暮色渐沉，沉落于 Domaine d'O 传统文化公园的橄榄林中，林中蝉鸣声悠远。这些昆虫在声学上的天赋，可谓众所周知：它们通过先进的腹部发音器，产生共鸣，并且能发出达 120 分贝的声音。而 Jean-Claude Carriere 剧场内却是安安静静，从外到内，双层木质墙体非常有效地阻隔了外界的声音。这个剧场位于在蒙彼利埃的 Domain d'O 公园中央，这是一片的 23 公顷的文化绿洲，旨在服务于艺术展、节日和活动。剧院是以当地一位很有声望的剧作家命名，是这个公园建设的最新一笔：一幢木建筑，这幢木建筑在松树、橄榄树和那些为了周末野餐设计的小庭院之间，如一个公园中升起的彩色立方体一般。

对于路过的人来说，首先吸引他们的就是建筑物的立面，那些富有趣味性的图案由上百个菱形组成，通过长条形的厚木板交错形成，布满了整个墙面。在一些菱形中，闪烁着不同颜色的灯。"我们有了这个想法，就是让这种传统的丑角戏服[1]的主题成为一个贯穿整个剧场的主题，所以我们想了很多不同的方式应用这一概念。最后，对于剧场的立面，我们决定创造一种菱形为基础的木质网格。" A+ 建筑的首席建筑设计师菲利普博昂是这样介绍的。同样，这个图案也一直在剧场内部重复出现。穿过小小的入口雨篷，给人带来一种从一块巨大毯子下面偷偷潜入的印象，当你进入到剧场的世界，更多的菱形给人带来视觉、触觉、感觉的冲击。明亮而通风的大厅被菱形的窗户所主导，创造了一种有吸引力的光与影的交织，这种氛围会随着日光的转移而变化。剧场的这种美学属性得到了访客的赏识。"这座建筑应该散发着光和想象力，这是一个任何事情都可能发生的地方。我们对于访客的评价非常欣慰。他们说，当他们在大厅坐在他们的位置上的时候，他们非常享受这里的氛围和木质的温暖香气。对于我们来说，访客能够欣赏礼堂声学效果是非常重要的，但是对于建筑和美学上的欣赏对我们来说同样重要。"

Jean-Claude Carrier 剧院是 A+ 建筑公司第一个完全使用木材设计的建筑。选择木材最初来源于客户的首要需求——这个建筑物会成为法国的第一个生态剧院。"通常来说，一个这种规模的项目将会考虑 30-50 年的能源用量。我们已经意识到建筑中的这个参量，并希望大幅度地提高能源效率。所以我们决定只采用可持续的和高效能的材料，从这点出发，引导我们在整个项目中使用了木材，在选择其他材料——橡胶地板、涂料和玻璃上我们也以同样的标准挑选。根据我们的计算，我们的项目只需七年[2]来涵盖它的能源用量。"这个剧院拥有各种杰出的的生态凭证：优越的保暖功能、智能的发热系统和低耗能 LED 灯的使用（包括舞台），这个建筑物是一个卓越的高能效的整体。从屋顶进入的充沛的自然光节省了电能。材料与充满活力的气候之间的互动也被充分地利用了。可以开合的屋顶减少了对空调的需求。但是，对于 1800 名（1200 座，600 站位）的观众容量，还有舞台上的演员们和聚光灯（即使使用的是相对来说散热较少的 LED 灯），剧场仍需用风扇来调节室温。当地开发商希望这个建筑物可以在需要的时候进行拆卸和迁移。这也是剧院远景的一部分——在未来可以把文化传播到蒙彼利埃以外的地方，这也是采用木材的一个原因。如果这个建筑物需要进行拆卸和迁移，那么这个过程对于周边环境的影响同样是需要考虑到的。预制木板的使用减少了建造时间和对工地周围环境的影响。总的来说，从一个想法到剧院的开幕，只用了一年时间。

Jean-Claude Carriere 剧院位于一个有 300 年历史的古老公园里，有着深厚的文化底蕴。在规划这座建筑时，这一方面也被自然的考虑其中。"我们的灵感来源于这个公园和戏剧的世界，两者都有强大的个性和悠久的历史。正因如此，我们希望我们的建筑能与二者都产生对话。建筑采用的红色是剧院的传统色调，与外墙的菱形图案相结合，立刻给人戏剧的联想。

剧院周围的松树林也给予了灵感！那是一片宁静的绿洲，是个能放松身心的地方。树林和木制的剧院相互辉映、互相提升，让二者都更加美丽。"剧院独特的红色源于一种

---

[1] 丑角的戏服原本是打着很多彩色补丁的农民的衬衫和长裤，后来它转变成了一种用三角形和菱形装饰的紧身戏。
[2] 我个人并不理解这个几年的能源用量的涵盖，通过资料查询我并没有找到相关的内容，如果有对建筑节能比较熟悉的人，希望能进行最后矫正。

Diamond acoustics

Angled wall panels damp echoes. A total of 1200 cubic metres of wood has been used in France's first ecological theatre, which sports its own Harlequin costume. Why? To create superb acoustics and give visitors the theatre vibe in an instant.

Text Erik Bredhe

AS DUSK SLOWLY descends over the olive groves of the classic culture park Domaine d'O, the buzz of the cicadas gets louder. The acoustic talents of the insects are well known and impossible to avoid. Equipped with an advanced resonance apparatus on the stomach, they create a dense wall of sound reaching up to 120 decibels.

Inside the Théâtre Jean-Claude Carrière, however, all is quiet. Sound from outside – and inside – is effectively damped by the theatre's insulated double wooden walls. The French theatre stands in the middle of the Domaine d'O park, in Montpellier's 23 hectare cultural oasis for art exhibitions, festivals and events. The Théâtre Jean-Claude Carrière, which is named after a famous screenwriter born in the area, is the park's latest addition, built entirely in wood. The theatre rises up like a colourful cube in the middle of the park among the pines, olive trees and pleasant small gardens that are just made for a Sunday picnic.

The first thing to strike passers-by is the theatre's facade, with its playful patterns of hundreds of rhombuses – diamonds if you will – created using long criss-crossing planks of wood, that climb up all the walls of the building. In some of the rhombuses, lamps shine in different colours.

"We had an idea that the pattern of the classic Harlequin costume should be a theme running throughout the theatre. So we thought about different ways of applying that concept. In the end, we decided to create a sort of diamond-shaped wooden lattice on the theatre's facade," explains Philippe Bonon, one of the lead architects at A+ Architecture.

The pattern is also repeated inside the theatre. As you walk through the entrance which, with its small protruding canopy roofs gives the impression of ducking under an enormous blanket, into the world of theatre you encounter yet more diamond patterns. The light and airy lobby is dominated by rhombus shaped windows that create an attractive play of light and shade – an effect that changes through the day as the sun moves across the sky. The aesthetic properties of the theatre are much appreciated by its visitors.

"The building should exude lightness and imagination, a place where anything can happen. We were so pleased when we heard the comments from the first visitors. They said they enjoyed the atmosphere and the warm scent of wood, in the lobby and when they went and took their seats. It's naturally important for us that the visitors appreciate the acoustic properties of the auditorium, but also its architecture and aesthetic."

The Théâtre Jean-Claude Carrière is the first building that A+ Architecture has designed entirely in wood. The choice of wood came primarily from the top priority of their client – the building was to be France's first eco-theatre.

"Normally a project on this scale will take between 30 and 50 years to cover its energy use. We were conscious of this parameter and wanted to significantly improve on it. We therefore decided only to use sustainable and energy-efficient materials, which is what led us to build the entire theatre in wood, and to choose the other materials – rubber flooring, paints and glass – with equal care. According to our calculations, our project will take just seven years to cover its energy use."

THE THEATRE HAS all sorts of standout eco-credentials. With its excellent insulation, a smart heating system and its use of low-energy LED lamps throughout – including on stage – the building is a highly energy-efficient creation. Electricity is also saved by allowing plenty of light in through the roof. The material's interplay with the dynamic Mediterranean climate has also been exploited. Having a roof that can be opened and closed reduces the need for air conditioning. However, with capacity for an audience of 1800 people (1200 standing and 600 seated) and with actors and spotlights on the stage (even if they are LED lamps that generate less heat than others), fans are required to regulate the temperature.

Another stipulation from the client, the local council – Conseil Général de l'Hérault, was that the building could be dismantled and relocated if required. This was part of the vision for the theatre to be able to disseminate culture even beyond Montpellier in the future. This became another reason for building in wood. If it was going to be possible to assemble and disassemble the building, thought also needed to be given to the impact that process would have on the surroundings. Using prefabricated wooden panels reduced both the construction time and the impact on the environment around the construction site. In all, from idea to opening, the process took just one year.

THE THÉÂTRE JEAN-CLAUDE CARRIÈRE stands in a 300 year-old park where culture has always played an important role. When planning the theatre, it was only natural to take this into account.

钻石声效

"We were inspired by the park and the world of theatre, both of which have a strong identity and a long history. We therefore wanted our building to be in dialogue with them both. The red colour of the building is a classic theatre colour, and it combines with the rhombus pattern outside to provoke immediate theatrical associations. The pine forest in which the theatre sits also gave us ideas! It's a wonderfully tranquil oasis, a place to free your mind. The forest and the wooden theatre lift and augment each other – making them both even more beautiful."

The theatre's characteristic red colour comes from the FunderMax panelling, a durable compact laminate. The wooden lattices that cover much of the facade are made from untreated French larch. The theatre's frame, roof trusses and structural elements are made from glulam beams. The interior of the roof and walls is then clad in cross-laminated timber panels. The ceiling is lined with a moisture barrier to protect the wood panels from rising moisture emanating from inside the theatre. All the interior wood is untreated, so that it can age naturally. Only certain details such as skirting, wet rooms and toilets have been given a coat of varnish, a surface treatment that will need to be repeated on a regular basis.

All the wood in the building – around 1200 cubic metres – is PEFC certified, which means the timber has been harvested and processed in line with internationally established standards that guarantee sustainable forestry. The theme of sustainability runs through the whole project, not just in the choice of materials, but throughout the life cycle of the building, including transport to and from the construction site, for example.

Wood was also important in achieving the right acoustic properties for the theatre. Details in the auditorium that are there to improve the acoustics in the hall even become part of the scenography.

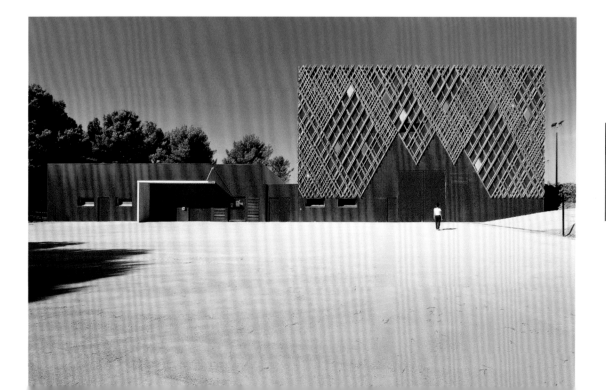

紧实耐用板。木制格子状外墙是由未经加工的法式落叶松制作而成。剧院的框架、屋架及其他结构的组成部分都是由胶合木梁建造的。室内屋顶及墙壁有交叉层压的木板作为覆层。天花板有一层防潮的涂料，避免木材被剧场里散发的湿气所侵蚀。所有用于室内的木材都是未经加工处理的，这样它们可以自然老化。只有一些细节部分，比如说踢脚、潮湿的房间以及卫生间都刷上一层清漆，并需要定期的再加工。

建筑物里所有 1200 m^3 左右的木材都是由泛欧洲森林认证体系 (PEFC) 许可的，也就是说，木材的采集和加工都已根据国际规范的标准，保证了绿色生态的可持续性。可持久性是这个企划的中心之一，不仅仅体现在选材上，也体现在了建筑物的生命周期，包括往返施工场地的运输等等。木材为给剧院实现良好的音响效果中起到了很重要的作用。会堂里为了提升听觉效果而添加的细节也成为了舞台美学的一部

分。"我们的设计使所有部分都有其功能所在,并体现这个建筑物的理念。我们想让它们以一种看得见的形式连接在一起。我们将会堂里的墙都倾斜在了一个特定的角度,由此可让演出时所有的声音和动作被清晰地传达,同时减少不必要的噪音。这么做的结果是,剧院具有了巨大的潜力——它可以承接多种的文化及艺术活动。"

因为这是一个公共建筑,它在法国被归类于更多的开放给公众的机构(Establishment Receiving Public,ERP),因此必须遵照有关于这类建筑物的严格规章制度,这些制度影响了对于本剧院的设计,比如说走廊的宽度和安全出口的数目及地点。但是,一个木结构的建筑物与钢结构的建筑物相比,这些制度要宽松一些。因为众所周知,木构建筑的耐热性更好一些。"承重性的结构以及墙体上结实的面板的墙体都提供了一个自然的防火墙,归功于它们巨大尺寸,所以不需要更多的加工。"Philippe Bonon 说。A+ 建筑设计工作室对于完成的结果非常满意。唯一的遗憾是建筑师们会非常想念建筑工地。"这是我们工作室史上首次运用木头到了这种程度。整个过程教给了我们许多关于材料的知识——当然也包括建筑设计!举个例子说,我们现在知道纯粹使用木材的建筑工地是什么样的。那是一个又干净又令人愉悦的工作场所。我们在那儿干了一年,每一天都是,所以我们会非常地想念那里。如果有这个机会的话,我们很期待在将来继续设计木建筑。"Philippe Bonon 总结道。

PEFC

泛欧洲森林认证体系是一个国际的非政府机构,致力于推广森林生态的可持续性。PEFC 的认证保证了木材是来源于根据可持续性的需求而管理的森林。

Diamond acoustics

A+ ARCHITECTURE

在拥有优秀的建筑师团队之外,这个位于巴黎的事务所也聘请了设计师、平面设计师和规划师。Jean-Claude Carrier 剧院是他们参与的第一个木建筑工程。A+ Architecture 的其他优异作品包括大型体育场。其中一个是在 Montpellier 市的 Park&Suits 竞技场,其可容纳 14000 人,这个竞技场是为了举办大型体育活动以及表演所设计的。

PEFC

PEFC (Programme for the Endorsement of Forest Certification) is an international non-governmental organisation that works to promote sustainable forestry. PEFC certification is a guarantee that the wood comes from sustainably managed forests.

A+ ARCHITECTURE

Along with its team of architects, the Paris-based firm also employs designers, graphic designers and urban planners. The Théâtre Jean-Claude Carrière is the first project they have created entirely in wood. Other standout projects in the A+ Architecture portfolio include large sports arenas. One of the largest is the Park&Suits Arena in Montpellier, with its capacity for 14,000 people. It is designed to host both major sports events and arena concerts.

"We designed everything to have a function, and to follow the architectural concept of the building. We wanted it all to link together in a very visible way. We fitted the walls inside the auditorium at a specific angle that allows every sound and movement to be heard during the performances and reduces unwelcome echoes. The result is a theatre with huge potential, which can host all sorts of cultural and artistic activities."

SINCE THIS IS A PUBLIC BUILDING, it is classified in France as an Establishment Receiving Public (ERP) and must therefore follow the stringent rules concerning this category of building. This affects design aspects such as corridor width and number of emergency exits, as well as the location of these. The criteria are, however, slightly less strict for this type of building, constructed around a wooden frame, than for buildings with a steel structure. This is because wooden constructions are known to tolerate heat better and remain standing for longer in the event of a fire.

"The load-bearing structure and substantial panelling on the walls offer natural fireproofing, thanks to their sheer size, and so did not need to be treated," says Philippe Bonon.

A+ Architecture is pleased with the finished result. The only negative aspect the architects can come up with is that they will miss the construction site.

"This is the first time in our firm's history that we've used wood to such an extent. It's taught us so much about the material – and about architecture! For instance, we now know what a site is like when working exclusively with wood. It's a very clean and pleasant place to work. We did it for a whole year, every day, and we're going to miss it a lot. If we get the chance, we'd very much like to work even more with wood in the future," concludes Philippe Bonon.

钻石声效

Libeskind in Berlin

李伯斯金在柏林

Text: Leo Gullbring | 校译：高瑜　韩佳纹　| 柏林犹太博物馆，德国，丹尼尔-李伯斯金（Daniel Libeskind）
摄影：BitterBredt & Linus Lintner | 2013年3月11日出版
| The Jewish Museum in Berlin, Germany by Daniel Libeskind
| Published 11 March 2013

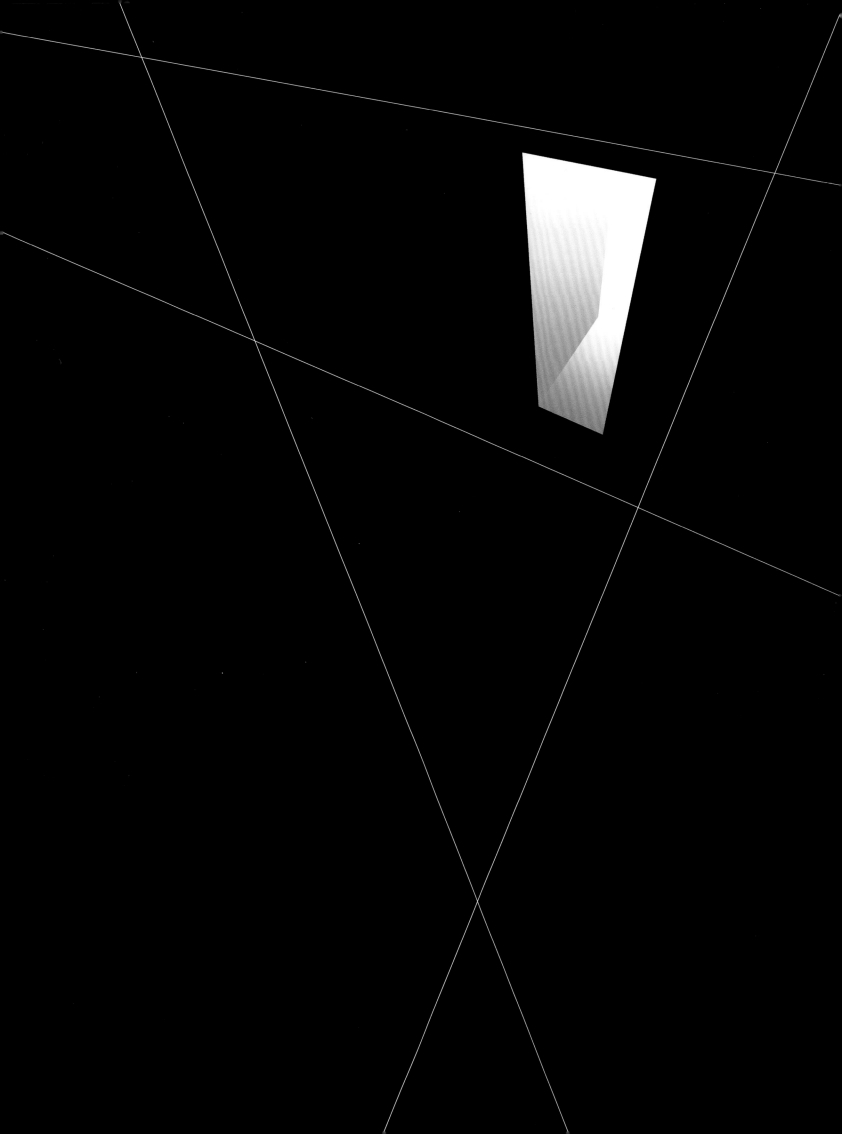

The Jewish Museum Berlin is expanding. When it opened in 2001, Daniel Libeskind's Jewish Museum Berlin sparked an international revolution in architecture. Now the museum is set to gain a new addition. Libeskind is continuing with the same deconstructed forms as before, but this time in three large, inclined cubes made entirely from wood – with allusions to Noah's Ark.

Text Leo Gullbring

THE JEWISH MUSEUM'S new Academy is being built right across the street, on the site of what was once Blumengrossmarkt, Berlin's flower market. The three crate-like entities in wood all have different geometric shapes, on a theme also found in the Garden of Exile and the Glass Courtyard that was added six years ago. However, the project is about more than just shape; it also makes a cultural and social statement. The Academy is something of a city in miniature, binding together the history of Berlin and the history of the Jews.

"We've used wood because it is beautiful, intimate and humane. Wood symbolises something organic and the building is like Noah's Ark, filled with knowledge. These units look like huge crates and they symbolise a piece of history that is returning to Berlin – they are more objects than buildings. We have striven to create space between the units and interlink the various activities of the Academy. We have also scaled things down for the benefit of all the children who come here," explains Daniel Libeskind.

The skylights illuminating the Academy's wood-panelled entrance cube are inspired by the two Hebrew letters Alef and Bet. They are also a reminder that learning and knowledge are important aspects of Jewish culture. Beyond the entrance lies a hall containing two more, larger cubes that house the library and an auditorium. Here too the timber-clad walls shoot in different directions, energising the in-between spaces, in line with the idea of 'Zwischenräume', Libeskind's title for the annex project.

THE ACADEMY'S ROOMS are made from roughly cut pine. The three cubes were prefabricated, reducing the construction work to just a few days.

Jorg Enseleit, designer at the engineering firm GSE in Berlin, explains that wood was chosen to achieve good spans with a lightweight design.

"All of the cubes are insulated because the old hall is not heated. The roofs of the interior cubes comprise boxes made from Kerto, a laminated veneer lumber, glued together and filled with insulation, while the exterior cube has TT-shaped Kerto elements with a similar solution."

The cubes' foundation plates are also insulated. Jorg points out that the geometry of the exterior cube is extremely complex,

柏林犹太博物馆正在扩建。2001年开馆伊始，丹尼尔－李伯斯金创作的柏林犹太博物馆引发了一次震惊国际建筑界的革命。当下这个博物馆正在准备扩建。李伯斯金将继续延续之前的解构形式，但这次采用全木结构打造3个大型倾斜立方——受启发于诺亚方舟的典故。犹太博物馆的新学院正在对街修建，即原来柏林花卉市场的地址。这三个箱状的木结构实体各自拥有不同的几何形态，6年前的流亡花园和玻璃庭院中也运用过这个主题。但是，这个项目的重点不仅仅在于其形态，它也是一个文化和社会的宣言。这座学院也可以看作城市的微缩体，连接起柏林的历史和犹太人的过去。

"我们采用木结构，因为它美观又具有亲和力。木材象征天然有机的物体，这个建筑也像载满知识的诺亚方舟。这些建筑单体看起来像一个个巨大的板条箱，它们象征着一段柏林历史的回归——它们不仅仅是建筑而是负有象征的实体。我们努力创造单体之间的空间，将众多的学院活动连接起来。我们也将物品的体量缩小以满足所有小朋友的到来。"丹尼尔－李伯斯（Daniel Libeskind）金解释道。天光点亮了建筑的木结构中的口，它的灵感来源于2个希伯来字母 א 和 ב。那也暗示着学识是犹太文化中极其重要的方面。在入口之上有另外两个更大的立方体，一个是图书馆，另一个为大礼堂。同样，木结构墙面朝不同方向延展拉伸，为过渡空间提供支撑，这也符合李伯斯金为新增建筑的命名"空间之间"。

学院的房间由粗略切割的松木搭成。3个立方体是预制构件组合的，工期缩短到仅仅数日。柏林建筑工程总承包商杰仕依集团的设计师 Jorg Enseleit 解释道，选择木材是因其拥有轻量化的设计和优越的跨度。"所有的立方体都是隔热的，因为旧的大厅没有供暖。立方体室内的顶部以 kerto 为材料制成的多个盒子构成，kerto 是一种单板层积材，黏合到一起之后再填充保温。外部由 TT 形状的 kerto 元件用类似的方法制造。"这些立方体的基础板也是隔热的。Jorg 指出外部立方体的几何形状是及其复杂的，那就是为什么钢铁被用作辅助材料。芬兰木业（FinnForest）生产的雷诺墙和 kerto 里帕面板（The Leno walls and Kerto-Ripa panels）被拧接在一起。特殊的钢板用于连接一些地方的屋顶和墙。其中两个立方体隐藏起来的，用蜡进行了处理，第三个立方体展露在外面，表面加覆了木材的保护层（Holzschutzcreme）。

Libeskind in Berlin

Libeskind in Berlin

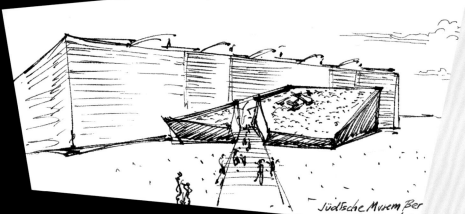

犹太博物馆每年有 7000 次博物馆游展和 4000 堂课程，内容囊括从孩子们的活动到博物馆职员培训等。新建筑将服务于会议、研讨会，以及所有的行政工作。从博物馆开馆以来，档案数量翻了一番，现在将有新空间对其进行合理分担了。青少年活动在学院中占很大比重，已经与克罗伊茨贝格区一个更大的学校签订协议。314 名学生中的绝大多数拥有穆斯林背景。除了引导关于伊斯兰和犹太教的讨论外，学院还会将问题聚焦在移民和差异性等。

柏林犹太教博物馆是欧洲最大的一个。开馆于 2001 年，那已经是李伯斯金在公开的建筑竞赛上排除万难赢得胜利之后 12 年。在他的太太及推广者尼娜－李伯斯金的帮助下，这个项目终于得以实现，尼娜曾在加拿大多次策划过竞选活动。他们一起创作这个博物馆，斜墙、又高又窄的房间、陡峭的楼梯和至少 1005 个窗户——每个大小都不一样。每扇窗都指向那一段独特的德国历史记忆。27 米高的房间中有一个狭窄的窗户，刺眼的日光照射进来，这源自一个女人的回忆，那是通向奥斯维辛集中营的火车：光在铁栅后闪烁，隧道里的灯，云和太阳的梦，以及希望能再次见到天光。

博物馆的建筑立面被锌包覆，镶嵌着扭曲的窗户，浑似那些被纳粹杀害者的地址簿。空旷的博物馆空间中狭窄人行道纵横交错，述说着柏林遗失灵魂的故事。这不单是解构历史和创造一种新历史观点的问题。这更倾向于打造一个空间来诉说长久以来被忽略的故事。犹太血统的丹尼尔理所当然是这项任务的最佳人选。他许多亲人都死于集中营：如果需要一个人有能力打造超越传统博物馆建筑的独特作品，那么他就是。这座建筑是出人意料的，它与弗兰克·盖瑞设计的西班牙毕尔巴鄂古根海姆博物馆一起，掀起了建筑上的国际革命。自从这两个博物馆开馆以来，全世界的建筑学家和设计师都被这种新的建筑风格大开眼界——建筑不仅是满足于功能的房子，还可以成为引发社会重新思索其灵魂的利器。

which is why steel was used as a supplementary material. The Leno walls and Kerto-Ripa panels come from FinnForest, and have been screwed together. Special steel plates have been used to join the roof and walls in some places. Two of the cubes are undercover and have been treated with wax, while the third stands outside and has instead been coated with a wood protector called Holzschutzcreme.

The Jewish Museum holds around 7000 museum tours per year and offers over 4000 courses, everything from children's activities to training for museum staff. The new building will provide conference and seminar rooms along with space for all the administration. The archives, which have doubled in size since the museum opened, will now have room to spread out properly. Activities for children and young people play an important role in the Academy, which has an agreement with one of the Kreuzberg district's large schools. Of the 314 students, the majority come from a Muslim background. In addition to leading discussions on Islam and Judaism, the Academy will also focus on issues of migration and diversity.

李伯斯金在柏林

The Jewish Museum Berlin is one of the largest in Europe. It opened in 2001, 12 years after Libeskind defied all the odds to win the open architectural competition. With the help of his wife and promoter Nina Libeskind, who in the past has masterminded election campaigns in Canada, the project was brought to fruition. Together they produced a museum with sloping walls, high narrow rooms, steep steps and not least 1005 windows – each one of a different size. All the windows refer to a unique memory drawn from the history of Berlin. A narrow window in a 27 metre-high room lets in stark daylight and derives from one woman's recollection of the train journey to Auschwitz. The light flickering behind a grille, the lights in the tunnels, dreams of clouds and the sun, and the hope of once again seeing daylight.

The exterior of the museum, which is clad in crisp zinc, was tattooed with these distorted windows, almost like an address book for every one of those murdered by the Nazis. Empty spaces in the museum are crisscrossed by narrow walkways and all of them tell the story of Berlin's lost soul. It was not so much a question of deconstructing history and creating a new historical perspective.

more about creating the space to tell a story that had long been neglected. Daniel's Jewish heritage was of course perfect for the task. Many of his relatives had perished in the concentration camps: If anyone was going to be able to create something unique that went beyond conventional museum architecture, it was him. The result was startling and, together with Frank Gehry's Guggenheim Museum Bilbao in Spain, it sparked an international revolution in architecture. Since the opening of these two museums, architects and designers around the world have had their eyes opened to a new type of architecture – one that not only fulfils a function as a building, but that is also an instrument for society to rediscover its soul.

It took a while, however, for an exhibition to be created, since the management was unable to agree on who should lead the museum. Nevertheless, in its first two years the museum still enjoyed record visitor numbers, since the architecture itself told the story. The architecture in many respects goes beyond the exhibited objects and has managed to make the invisible visible. With his building, Libeskind consigned the neutral white museum boxes of Modernism to the scrapheap of history and paved the way for architecture that dares to embrace more than technical functionalism. The architecture aspires to be an artwork of its own that reflects and inspires contemporary culture.

DANIEL READJUSTS HIS horn-rimmed glasses and stresses that the emotional is just as important as the structural. The experience becomes the focus when architecture is more about inclusion and expression than providing a static structure.

但是展陈的设计略花了一些时间，因为对什么主导博物馆的问题一直不能达成共识。然而，在最初两年里，博物馆游客数量一直在创造纪录，因为建筑本身就在讲述故事。建筑本身在很多方面已经超越了展品，并使得无形现于眼前。在他的建筑上，李伯斯金将现代主义的中性白色博物馆盒子束于历史之高阁，为敢于超越技术功能主义建筑的作品铺平道路。这座建筑渴求自身就成为艺术品，反映并启发当代文化。丹尼尔扶了扶他的牛角框眼镜，强调说情绪的和结构的同样重要。当建筑更多的是包容和表达而不仅仅是静态的结构的时候，体验感就会成为焦点。"每一个人都会对生活中的大问题感兴趣。如果建筑不能触动这样得问题，那就不能名副其实。对我而言，建筑本质上是与道德有关的。太多的建筑师缺乏信仰，并且毫无责任感地盲目遵从形式的条条框框。他们没有意识到他们所建造的会影响人们的生活。"

李伯斯金在柏林
Libeskind in Berlin

李伯斯金许多年没有在柏林工作，现在的工作室设在曼哈顿城区的雷克特街。他可以直接望到华尔街和三一教堂的尖塔，随时都在那些建于20世纪初期高耸入云的白色摩天大厦之外。摩天大楼风靡的初期，建筑师们都直接从样本上照搬外墙。整个欧洲历史文化都在摩天楼之中展示，从文艺复兴宫殿、希腊神殿、新艺术运动，甚至包括《卡里加里博士的小屋》的表现主义。随后，现代主义基于其方盒子的创造把建筑装饰中释放出来。李伯斯金自己的建筑曾是国际风格的反对者，也并未对其有任何叙述，但现在他更愿意讲起这些建筑之间街道上的故事，而不是曾经让他风靡的解构主义。"世界是在变化的，所有旧的意识形态都面临挑战。建筑是一种经验主义的手艺。但你看它的时候，你可以在日常生活中找到真谛并发现人们如何在社会中给自己定位。我的意思是：什么是艺术，什么是建筑？难道真的有一个论点可以试图去诠释'建筑'这个词吗？"

"Everyone is interested in life's big questions. If architecture can't touch on such issues, it is not worthy of the name. For me, architecture is essentially about ethics. Far too many architects lack belief and instead slavishly follow norms and rules without taking responsibility. They fail to realise that what they build affects people's lives."

After countless years working out of Berlin, Libeskind now has his office on Rector Street in downtown Manhattan. He can look out towards Wall Street and the pointed spire of Trinity Church, and immediately outside the white skyscrapers of the early 20th century stretch high up towards the clouds. In the early days of the skyscraper, architects copied the facades from pattern books. The whole of Europe's cultural history through Renaissance palaces, Greek temples, Art Nouveau and even The Cabinet of Dr Caligari is paraded among all the lofty buildings. Modernism then freed architecture from all this ornamentation in its boxy creations. Libeskind's own architecture was a reaction against the international style of the day with its absence of narrative, but today he prefers to talk about life on the street between the buildings than about the Deconstructivism that once made him famous.

"The world is changing, all the old ideologies are being challenged. Architecture is an empirical craft. However you look at it, the truth can be found in everyday life and in how people find their place in society. I mean: what is art, what is architecture? Is there really any point in trying to define the word 'architecture'?"

Daniel Libeskind has designed several Jewish museums, including a temporary museum built entirely from wood in Copenhagen in 2004. 2003 saw him win the competition to design a new World Trade Center and he has recently taken on the design of a peace centre on an old prison site in Northern Ireland. When asked whether he sees himself as a Jewish architect and whether that brings a certain responsibility, he raises his eyebrows and gives a broad smile. He explains that of course he has a esponsibility because, like other architects, he designs buildings. But there is also a side of Jewishness that is about discussing life's big questions. It is not just about ideals and fetishising objects, but more about discussing what is important in life. Impressive and beautiful facades are all very well, but there has to be some content there as well.

Libeskind in Berlin
李伯斯金在柏林

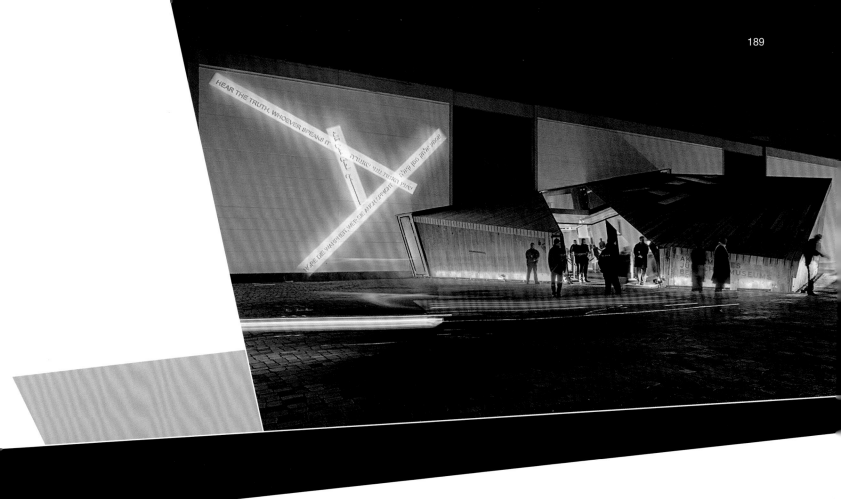

丹尼尔-李伯斯金已经设计了好几个犹太博物馆，包括一个2004年在哥本哈根完全用木结构修建的临时博物馆。2003年他在新世界贸易中心的设计比赛中获胜，最近他投入于北爱尔兰监狱旧址上一处和平中心的设计。当问到他是否将自己定位为犹太建筑师，是否带有特定的责任的时候，他抬起眉毛开怀大笑。他解释说他当然有责任，因为就像其他建筑师一样，他设计的是房子。但是这也关乎犹太特性的一个方面就是去讨论生活中的"大问题"。不仅仅关于完美的和蛊惑人心的事物，更多的是讨论什么才是生活中重要的事情。拥有使人印象深刻且美丽的外观非常好，但仍需要有同样分量的内涵与之匹配。

"我相信当下建筑最大的挑战关系于我们的记忆。可持续发展的房子当然重要，但现在也仅限于技术问题而已。一个真正的生态循环的解决方案是文化性的，那是我们心灵和灵魂的一部分，那关乎如何作为一个人的存在。不然我们就生活在一个阿尔茨海默症的世界里！记忆是人类基本的东西，记忆就是结构且对我而言建筑就像书籍，但是它们是用空间设计来书写的。建筑师需要有接受挑战和直面的勇气。卡夫卡写道，他从来不看他喜欢的书，只看那些困扰和激怒他，那些让他心中破冰的书！"

柏林犹太博物馆及其扩建都远离现代主义口号"形式追随功能"。相反，建筑自己就讲述故事，以激发想法和情绪。"新的学院和博物馆本身给柏林人一个机会去了解他们先人的生活，从而了解它们的城市，了解到大屠杀的后果不仅仅是亿万犹太人死亡也同样摧毁德国的文化。建筑不止是解决问题，也同样关乎提出问题。这在建筑中是少见的，但是在哲学、艺术、音乐中并不少见。为什么建筑不能解决存在的问题？它们关乎我们所有人，不仅仅是政治决策者：为什么我们在这里？我们将去向何方？有什么是凌驾于我们存在的社会秩序之上的？什么是生命的意义？

李伯斯金工作室

1989年，丹尼尔-李伯斯金在比赛中赢得了犹太博物馆的设计比赛，同时李伯斯金工作室成立于柏林，其他众所周知的作品包括英国曼彻斯特的帝国战争博物馆，美国丹佛艺术博物馆，德国奥斯纳布吕克的菲利克斯努斯鲍姆博物馆。

2003年，李伯斯金自由塔的设计赢得了曼哈顿下城发展公司主办的后9/11世贸遗址重建的比赛。现在只存留了他的设计，因为世界贸易中心的业主劳瑞-西尔弗斯坦另外任命SOM建筑设计事务所的大卫-查尔德斯设计来新的摩天大楼。

柏林犹太博物馆

柏林犹太博物馆于2001年9月开馆，是欧洲最大的犹太博物馆。开馆10年来，博物馆迎来超过700万的参观者。甚至在正式开馆前，博物馆建筑就吸引了35万参观者前来。新学院，占地2300 m²，图书馆和档案馆搬迁后便可以为旧馆腾出新的空间。

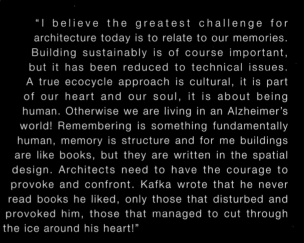

"I believe the greatest challenge for architecture today is to relate to our memories. Building sustainably is of course important, but it has been reduced to technical issues. A true ecocycle approach is cultural, it is part of our heart and our soul, it is about being human. Otherwise we are living in an Alzheimer's world! Remembering is something fundamentally human, memory is structure and for me buildings are like books, but they are written in the spatial design. Architects need to have the courage to provoke and confront. Kafka wrote that he never read books he liked, only those that disturbed and provoked him, those that managed to cut through the ice around his heart!"

The Jewish Museum Berlin and its new addition are far removed from the Modernist mantra 'Form Follows Function'. Instead it is an architecture that tells a story, provoking questions and emotions.

"The new Academy and the museum itself give Berliners a chance to get to know their city through the lives of their forefathers, through the Holocaust that not only brought death to millions of Jews but also destroyed German culture. Architecture is not all about solving problems, but about asking questions. This is rare in architecture, but not so in philosophy, art and music. Why can't architecture also tackle the existential questions? They touch us all, not just political decision-makers: Why are we here? Where are we going? What is there beyond he social order of our own existence? What is the meaning of life?

STUDIO LIBESKIND

Studio Libeskind began life in 1989 in Berlin when Daniel Libeskind won the competition to create a Jewish Museum on Lindenstrasse in Kreuzberg.

Other well known works include the Imperial War Museum North in Manchester, UK, Denver Art Museum, USA and the Felix Nussbaum Museum in Osnabrück, Germany.

In 2003 Libeskind's Freedom Tower design won the Lower Manhattan Development Corporation's competition to redevelop Ground Zero after the 9/11 attacks. All that remains now is his master plan, since the World Trade Center's owner Larry Silverstein instead commissioned David Childs from SOM to design the new skyscraper.

THE JEWISH MUSEUM BERLIN

The Jewish Museum Berlin opened in September 2001 as one of the largest in Europe. Over the 10 years since it opened, the museum has had over 7 million visitors. Even before the official inauguration, the museum building attracted 350,000 visitors. The new Academy, measuring 2300 m², will free up space in the old museum when the library and archives are moved across.

Folkhem's block
人民之家高层公寓

Text：Björn Ehrlemark ｜ 高瑜　韩佳纹
摄影：Rasmus Norlander

| Strandparken in Sundbyberg, Sweden
| Published 12 September 2013
| 松德比贝里，斯特瑞德帕肯，瑞士
| 2013 年 9 月 12 日发行

Showstopping showcases
Timber-framed high-rises are making a name for themselves in Europe. In recent years, six, seven and eight-storey timber buildings have risen up in Sweden and around the world. Come and explore Strandparken in Sundbyberg, Murray Grove in London and Via Cenni in Milan – three examples of how, following the latest advances, high-rise blocks have gone from experimental to mainstream.

Text Björn Ehrlemark
THE GENTLE SCENT from the cedar facade as it warms in the morning sun is expected. The wood block floor in the stairwell is a surprise. But the dizziness is breathtaking. From the duplex apartment below the pitched roof in the first of Folkhem's four eight-storey blocks in Strandparken, Sundbyberg, designed by Wingårdhs Arkitektkontor, the eye is drawn past a glass balcony, accelerates along 26 metres of shingled facade and continues straight down into Bällstaviken below.

People are currently moving into the building, which is the tallest timber-framed block in the Stockholm area. The environmental profile was always a key element of the project, right from the very start – the buildings were to be built in timber, and the wood was to be clearly on show. The facade is clad in knot-free cedar shingles, a choice that came after a great deal of debate about the right solution.

精彩案例展示
木结构高层建筑在欧洲开始扬名。近年，6层、7层、8层楼的木结构建筑已经在瑞典等全球各地建起。根据最新进展，在斯特瑞德帕肯的松德比贝里，英国伦敦的默里格罗夫和米兰森尼路这三个地方，可以找到高层建筑从小试牛刀走向主流之奥秘。

比约恩·俄勒马克撰文
清晨的阳光温暖了外立面，轻柔的雪松香味扑面而来。楼梯间的木地板是一处惊喜。但晕眩是令人激动的。在斯特瑞德帕肯的松德比贝里，由Wingårdhs Arkitektkontor 设计的四幢中第一幢"人民之家"八层建筑，坡屋顶下是复式公寓，目光会经过玻璃阳台沿着长达26米、覆满木瓦的外立面直降而下进入Bällstaviken 河，

人们现在正在搬入这幢斯德哥尔摩地区最高的木结构建筑。从一开始，环境状况一直都是该项目的关键因素，建筑要用木材建造，且木质要清晰地得到展示。外立面需被无瘤节雪松木瓦覆盖，这个选择经过了一大番关于正确方案的辩论。

"Resolving the details of a facade as large as this requires a whole new set of rules. It would perhaps seem obvious to go for the usual smooth cladding, but there's quite a difference between a summer cottage and a high-rise block," says Hanna Samuelson from Wingårdhs, who has worked on the project along with others such as lead architect Anna Höglund and engineer Anders Stålberg from the same office.

Hanna Samuelson lists some of the challenges they faced in erecting the building. What do you do about joints? Capillary effect? And maintenance? What tenant owners' association wants to be up oiling the exterior walls every five years?

"We have to take the wood seriously, it has to breathe, it absorbs water, it shrinks and swells and it moves," she explains, stressing that the material has to be seen differently from the design program's abstract, shiny surfaces that rise up to whatever height without any concerns.

The solution was cedar shingles, which are naturally resistant to rot and fungus and do not require treating. The trick is that all the small individual details give the facade a sort of pixelated effect that is less sensitive to variation. Each shingle can vary incredibly in colour – starting off dark red, or sometimes brown, and gradually turning grey – without the facade as a whole looking patchy or blotchy. The idea is that it will still look good 10, 15 or 30 years from now.

Folkhem's block

Although the structural frame is not exposed in the interior, surfaces in natural materials have been a guiding principle. The material palette for everything from ceilings to elevator interiors is made up entirely of carefully chosen woods. The internal window surrounds and the floor are ash, the balcony parapets are cedar and the window frames are stained pine. The wood can be both seen and touched, giving the new residents the natural feel they want. And they get good air quality into the bargain.

"No emissions from plastics and laminates or spores flying around. And the building is dry from the outset, so there is no integral damp to cause problems," states Hanna Samuelson.

ALTHOUGH BUILDING HIGH-RISE blocks in timber has been permitted in Sweden for almost 20 years now, questions from prospective buyers and visitors about the material choice in Strandparken often tend to come around to fire.

"I usually offer matches to all the visitors, so they can give it a try," says Arne Olsson, CEO of the developer Folkhem, as he points to the exposed timber frame in the showroom.

Setting fire to 170 millimetre cross-laminated timber is

"解决外立面的细节问题同需要一整套的新规则同样重要。这或许看起来像常规光滑覆面一样无奇，但是避暑别墅和高层建筑有很大差异。" Wingårdhs 建筑设计事务所的 Hanna Samuelson 说，她同该事务所的首席建筑师 Anna Höglund 和工程师 Anders Stålberg 等一起致力于此项目。

Hanna Samuelson 列举了一些在建设中面临的挑战。拼接处如何处理？毛细作用呢？怎么维护？哪种房东愿意让外墙每五年上一次油。

"我们必须谨慎对待木材，它需要呼吸，会吸水，会收缩膨胀，还会移动。"她解释到，并强调说材质不同于设计方案上抽象的光泽表面，上升到任何高度都不用顾虑。

解决方案即是雪松木瓦，既天然抗腐、抗菌也无需处理。有趣的是每一个小细节使得立面有类似于像素化的效果，但变化不会大。每一块木瓦都会神奇的在色彩上发生变化——从暗红色或类似棕色开始，渐渐变成灰色——而不是外立面整个变得残缺或者斑驳。计划是从现在开始往后 10 年、15 年甚至 30 年，都看起来不错。

尽管构架未在室内展现，但天然材质的外表已经成为指导性原则。从天花到电梯箱内的面材都是由精心挑选的木材打造。室内窗框和地板是白蜡木，阳台护栏是雪松，窗户结构是橡木。木材既看得见又摸得着，能够为乔迁者带去他们所期待的天然质感。他们也可以得到优良的空气质量。

"没有塑料、层压板的放射物或粉尘四处飞扬。建筑物从一开始就是干的，所以不会有湿气引发问题。" Hanna Samuelson 说。

尽管高层木结构建筑早在约 20 年前就在瑞典得到准建批示，但在斯特瑞德帕，潜在买家和来访者通常都在材质选择上问出关于火灾的问题。

"我经常会给所有的来访者提供火柴，请他们试试。"开发商"人民之家"的首席执行官 Arne Olsson 指着样品间暴露出的木结构说。

点燃 170 毫米直交层积材 (clt) 类似于试图点燃一根树干。Arne 指出公寓里有易燃危险的是室内家具，不管房子是什么造的。

对许多人来讲，是烧焦残余物的画面所带来的反射作用。也许相对于木材焚后的残片，钢梁在灼热中扭曲并不常见。与钢结构相反，木材在火灾中表现得更有预见性，一旦紧急情况

like trying to light up a tree trunk. Arne points out that what burns quickly and is dangerous in an apartment is the interior furnishings, no matter what the building is made from.

For many it is a reflex action conjured up by images of charred remains. Maybe this is because a steel girder twisted in the heat is an uncommon sight compared with a burning piece of wood. In contrast to steel, wood behaves predictably in a fire, which makes it easy to get the specifications right and safer for the fire service to go in if an emergency does arise. The firemen can quickly see more or less how long a timber-framed building will stay standing, even if it is on fire. The charred surface that forms at high temperatures on a load-bearing timber beam or post protects the material inside. The structural frame is therefore designed according to the length of time the building has to stand for under fire regulations. It must be possible for several centimetres to become charred without the structure losing its load-bearing capacity.

The apartments in Strandparken are fitted with domestic sprinklers, as an extra safety measure. In the event of a facade fire, the system cools both the rooms and the windowpanes to protect against any external heat. Wingårdhs architect Hanna Samuelson again:
"There's a lot of psychology in this. Everyone knows that wood burns. But then what happens to a facade of cellular plastic covered with a thin layer of render?"

THE CLT SURFACE UNITS from which the building is constructed were supplied by the Martinsons factory in Västerbotten. They are one storey high and as wide as one apartment.

"The risk with a prefab building is that it looks like a patchwork quilt with all its joints. The great thing about the shingles is that we can attach them to the wall units in advance and then sort out the horizontal and vertical joints once they are up," says Hanna Samuelson.

On closer inspection, the precision is impressive. Row after row of neat shingles wrap around the building. They line up with the windows, come out onto the balcony parapets and reach over the pitched roof, and in the corners they meet like the fingers on a pair of clasped hands – which is exactly as it should be. Developer Folkhem has always had high environmental ambitions for the project.
"What we thought was: If we put up timber-framed buildings and can make a name for ourselves as truly environmental builders in this city, we might be able to source land more easily and successfully. After all, isn't every politician interested in doing their bit on climate issues?" says Arne Olsson.
"A building like this stores as much carbon dioxide as 40 concrete buildings. That's why we see no need for any more debate on what construction materials to use."

THAT ATTITUDE HAS NOT been quite so prevalent in other high-rise timber projects, not even high-profile ones. A prime example is when architects at Waugh Thistleton tried to sell their design for a nine-storey block in solid timber in Murray Grove, London, to their client. They highlighted the environmental advantages of wood and the accompanying PR coup, but the argument fell on deaf ears. The fact that the material would reduce carbon emissions by over 300 tonnes compared with an equivalent building in concrete was not a viable argument in the property market, countered the client, because who wants to pay London prices to live in an 'experiment'? Then the architects explained that the Austrian solid timber frame could be delivered on about 20 lorries and erected at a rate of one storey per week. This was an argument the client could buy, but with one ultimatum: that the choice of material would be kept secret when the apartments were released onto the market.

The interiors were clad in plasterboard and the outside of the timber building was concealed behind a facade of fibre cement panels. However, news of the impressive timber design was leaked to the press and once published, all 29 apartments sold in just one and a half hours.

In a suburb of Milan, the cranes have recently left the continent's largest timber construction

site. The four residential high-rises were designed by RossiProdi Associati and are connected by two-storey terraces that form enclosed courtyards. Together they form the new housing complex Via Cenni.

The project is part of the reconstruction process after the massive earthquake in L'Aquila in 2009. The load-bearing walls, floor structure and elevator shafts of the buildings have been constructed in cross-laminated timber totalling 6,000 m. The choice of wood partly came down to a question of time, since it was important to replace damaged and condemned buildings across the region as quickly as possible.

Making sure the disaster would not be repeated was also an important priority when choosing wood as a construction material. The timber-framed design is estimated to weigh around a quarter of the equivalent building in concrete, which means that if some part of the building were to collapse, the damage and injuries would be less severe. Even more importantly, the lower mass reduces the forces that build up when the ground shakes, which makes the building much better protected against future earthquakes.

ON WAGRAMERSTRASSE IN VIENNA, Austria's tallest residential timber building has just been handed over to its tenants. The block is seven storeys high and was designed by Schluder Architektur.

At the same time, a new building is being constructed to house the Wood Innovation Design Centre in British Columbia, Canada, providing offices and public spaces. This one, by architects mga, is likely to be the world's tallest timber-framed design. But that is not likely to remain the case for long – even the architects hope the record will soon be beaten.

Perhaps that will happen in Bergen, Norway, where the green light has just been given for the residential tower 'Treet' which, at 14 storeys, will lay claim to the title of 'tallest timber-framed building in the world'. The project

发生，能方便正确规格的消防设施安全进入。消防员会很快弄清大概木结构建筑可以矗立多久，即使已经着火了。承重梁和承重柱在高温下形成的烧焦表面保护着内在的材质。因此，结构框架要根据消防法规下建筑可以支撑的时长来设计。必须允许在好几厘米烧焦的情况下，结构不会失去承重能力。

在斯特瑞德帕的公寓都配有家用喷淋装置，更多一份安全保障。如果外立面发生火灾，系统会使房间和窗玻璃都冷却，用以抵御任何外界高温。Wingårdhs 建筑师 Hanna Samuelson 又说："这里面有很多心理学。人人都知晓木材会燃烧。但是有薄薄粉刷层的泡沫塑料外立面又会怎样呢？"

构成楼房表面的直交层积材 (clt) 是由 Västerbotten 的 Martinsons 工厂提供的。它们有一层楼高，一个公寓那么宽。

"组合式建筑的风险在于建筑上的接缝使之看起来像百纳被。木瓦最棒之处在于我们可以提前将之固定到墙面组合单位上，然后再将水平接缝和垂直接缝分门别类。"Hanna Samuelson 说。

经过进一步观察，其精确度令人惊叹。一排排齐整的木瓦包裹着建筑。它们与窗户齐平，覆盖阳台护栏，延伸上斜屋顶，在每一个拐角处都如紧扣双手的十指一般——恰如其分。开发商"人民之家"一直对这个项目有很高的环保期待。

"我们计划的是：如果我们建造起木结构建筑，并且我们能在这个城市中以真正的环保建设者成名，我们会更容易地成功寻得土地。毕竟，不是每一位从政者都致力于为气候问题做出贡献吗？"Arne Olsson 说。

"这样的一个建筑可以储存 40 幢混凝土建筑才能储存的二氧化碳量。这就是为什么我们认为没有必要再讨论采用什么建筑材料。"

这种看法没有在其他高层木材项目中得到普及，更别说饱受关注的项目了。有一个典型案例，当 Waugh Thistleton 事务所的建筑师试图将位于伦敦默里格罗夫的九层实木建筑设计卖给客户的时候。它们强调木材的环保优势和相关性能要求的妙处，但观点并未得到采纳。此种材料的建筑可以比同等的混凝土建筑减少 300 吨的碳排放量，这在房地产市场中并不能成为一个可行的论据去说服客户，原因是谁愿意用伦敦的房价去入住一项"实验"？后来建筑师解释到，可以确保提供约 20 卡车的奥地利实木结构材料，并且以每星期一层楼的速度进行建设。这成为客户愿意买单的理由，但是有一项需无条件接受的要求：公寓在市场投放中，材料的选择将被保密。

室内被石膏板覆盖，木建筑的外表被隐藏在纤维水泥板的外立面内。但是，当精彩的木构设计被泄露给媒体并公之于众，所有 29 套公寓在一个半小时内售罄。

在米兰的一个郊区，最近起重机为这片大陆留下了最大的木建筑工地。这 4 幢高层住宅是由 RossiProdi Associati 操刀设计，与两层楼的露台相连形成封闭庭院。它们共同组成森尼街道的住宅综合体。

这个项目是 2009 年拉奎拉大地震后重建过程的一部分。承重墙、地板架构和电梯井由共计 6000 米长的直交层积材 (clt) 建造。木材的选择一定程度上是因为时间原因而定下来的，因为尽可能迅速地替换整个地区受损和废弃的房屋很重要。

确保灾难不会重演也是选择木材为建筑材料的一个重要先决条件。木结构设计大约只有同等混凝土建筑四分之一的重量，也就是说就算建筑的一些部分坍塌了，破坏和伤亡都不会那么严重。甚至更重要的是，当地面晃动的时候，更低的重量会降低建筑所产生的力量，从而使建筑能更好地防御未来的地震。

在维也纳的 WAGRAMERSTRASSE 路上，奥地利最高的木结构住宅楼刚刚移交给住户。这幢楼有 7 层楼高，是由 Schluder 建筑设计事务所设计。

人民之家高层公寓
Folkhem's block

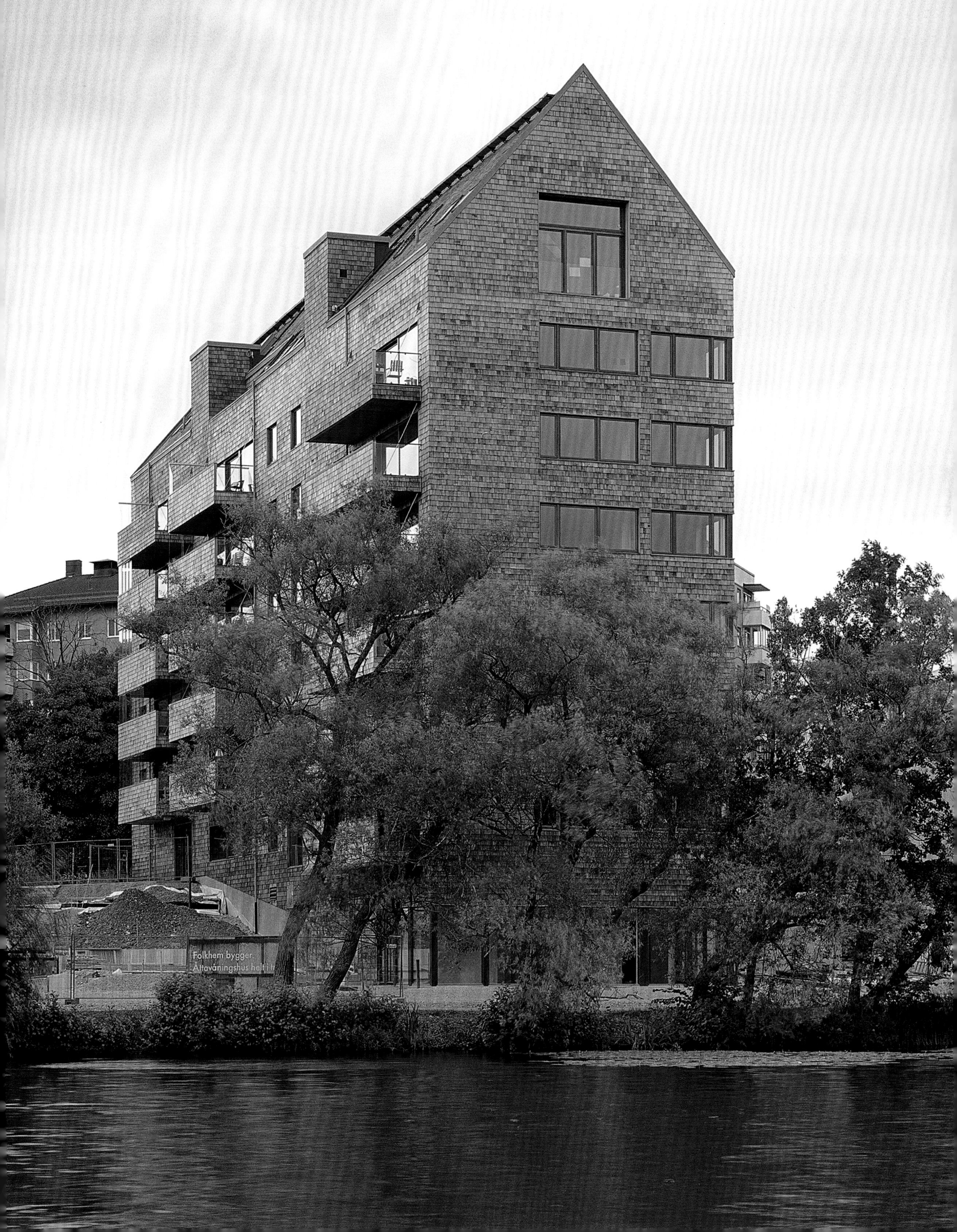
Folkhem bygger.
Åttavåningshus helt i

was designed by Marina Trifkovi and Per Reigstad from the Artec firm of architects and is described as groundbreaking with its combination of load-bearing glulam and modules. It will be 47 metres tall and should be ready to move into by early autumn 2014.

The US architectural giant Skidmore, Owings & Merrill, which designed the Burj Khalifa and five more of the world's 12 tallest buildings, also recently published a research report on a system for timber skyscrapers. The report describes the timber construction technique as an evolution of the formula from the past century for tall office blocks in steel and glass that once set the standard. And London-based Arup, often referred to as the world's leading consulting engineers, have also just tackled the issue of tall timber buildings with a fresh high-rise concept called the Life Cycle Tower.

100 years ago, 10 floors of cast iron or concrete would scare passers-by, who were convinced the building was going to fall down on them, but high-rises soon became an established part of the cityscape. Similarly, timber high-rises are becoming more accepted as they become more common. As the push for altitude escalates, more and more applications are also coming to light.

BACK AT STRANDPARKEN in Sundbyberg, a removals van has just pulled up outside the door. Storage boxes share the elevator with tradesmen on their way up to put the finishing touches to the interior in the duplex apartment in the pitched roof. Before the final finishes are applied, it is still possible to see how all the installations fit into the floor structure and how the heating coils are channelled under the floor. No radiators, so the large, sliding patio doors go all the way down to the floor.

与此同时，在加拿大不列颠哥伦比亚省正在为"木结构创意设计中心"修建一处带有办公空间和公共空间的新建筑。Mga建筑设计事务所表示这将成为世界上最高的木结构设计。但这也不会保持很久——就连建筑师也希望这项纪录能很快被打破。

或许那将发生在挪威卑尔根，14层的住宅高层treet刚刚得到批示，宣称将成为世界上最高的木结构建筑。这个项目由Artec建筑公司的Marina Trifkovi和Per Reigstad设计，将承重胶合木与模数相结合被描述为具有开创性的意义。建筑将有47米高，计划于2014年初秋接受入住。

美国建筑设计巨头Skidmore, Owings & Merrill事务所，曾设计迪拜塔和世界上最高的12幢大厦中其他5幢，最近也发表了木结构摩天大厦系统的研究报告。报告将木结构建筑技术描述为上世纪模式的进化，一度钢结构和玻璃幕墙成为高层办公大楼的标准模式。总部位于伦敦的Arup，一直被称为世界咨询工程的龙头，也采用了被称为"生命周期塔"的高层新概念来应对高层木结构建筑的问题。

一百年前，10层楼的铸铁或混凝土会把路过者吓住，他们觉得房子会倒在他们身上，但很快高层建筑成为城市景观的既定部分。同样地，木结构高层也会随着之普及而越来越被接受。随着高度的快速增长，越来越多的用途也将逐渐为人所知。

回到松德比贝里的斯特瑞德帕肯，一辆搬家卡车刚刚在门外停下。储物盒与将最后内装拿进斜屋顶下复式公寓的技术工人共乘电梯上楼。在封板之前，仍可以看到所有设备是如何完美装配到楼板结构中，加热盘管是如何在楼板下蜿蜒的。没有暖气，所以大大的中庭滑门一直往下到地板。

For optimum sound insulation, the ceiling is suspended from the walls, not from the floor structure above, and the walls separating the apartments are made from double solid wood panels separated by a 20 mm air gap. Each apartment becomes a separate acoustic box.

Folkhem's tall, thin block is the first of a group of seven to be completed, all of which will have one gable overlooking the waterside promenade along Bällstaviken and the other facing Sundbyberg's Hamngatan. The four middle blocks will be timber buildings with a similar design to the first one. These will be flanked by prefabricated concrete buildings constructed by another developer.

New residents therefore have a building site on either side of them, but they are each quite different: From the north side comes the screaming of a grinder on a concrete section and its reinforcement bars. On the south side, a huge white marquee stands ready to follow the next new timber building, step by step, as it grows up from the ground. Inside the marquee it is quiet and dry.

"And it smells of freshly sawn timber," says Hanna Samuelson.

TIMBER HIGH-RISES
The race is on for the world's tallest timber building, with entrants spread around the globe. Leading lights such as Canada's mga, Norway's Helen & Hard and London-based Waugh Thistleton have been joined by big names such as Skidmore, Owings & Merrill and Arup. The latter two have both recently launched projects outlining the potential to build up to 30 storeys high – true timber skyscrapers.

In Sweden, the first of four eight-storey buildings with a solid timber frame and a cedar shingle facade has just been completed on Bällstaviken in Sundbyberg. Designed by Wingårdhs Arkitektkontor for Folkhem, this is the first timber high-rise in the Stockholm area. But it is also part of a long-term focus on the part of the developer.

Folkhem's block

为达到最佳隔音效果，天花是从墙面悬吊下来的，而不是上层楼板架构，隔墙由气隙 20 毫米的双层实木板材制成。每户公寓都是一个独立的声音承载空间。

"人民之家"高高瘦瘦的建筑是一组七幢里面第一个完工的，每幢都有一个山墙俯瞰沿 Bällstaviken 的亲水长廊，另一个朝向松德比贝里的哈门街。中间四幢房子会是跟第一幢类似设计的木结构建筑。它们侧面将会由其他开发商建造预制混凝土房屋。

新居民因此可以见到一边一个建筑工地，但两者完全不一样：北边是磨床在混凝土块和钢筋上发出的尖锐刺耳的声音。而南边，一个巨大的白色帐篷时刻准备跟随下一个木结构建筑一起渐渐拔地而起。帐篷里面是安静且干燥的。

"闻起来是新鲜的锯材味道。" Hanna Samuelson 说。

木结构高层

世界最高的木结构建筑的竞赛已经拉开，参赛者遍布全球。领跑者包括加拿大的 mga，挪威的 Helen & Hard，总部位于伦敦的 Waugh Thistleton，还有众所周知的 Skidmore, Owings & Merrill 和 Arup。后两者都于近日启动项目，概述建造 30 层楼高木结构摩天大厦的可能性。

在瑞典，四幢用实木框架和雪松木瓦立面建造的 8 层建筑中第一幢已经刚刚在松德比贝里的 Bällstaviken 竣工。由"人民之家"的 Wingårdhs Arkitektkontor 设计，这是斯德哥尔摩地区的第一幢木结构高层建筑。这也是这类开发商长期关注的部分。

城市木业进展
Urban development in wood

Text: Sture Samuelsson | 校译：倪雯
摄影：Jack Hobhouse,David Valldeby,Morten Pedersen/Inviso,Mikko Auerniitty,Oopeaa,Bernd Borschart,Hawkins\Brown

in Norway, Finland, Germany and UK by Artec, Oopeaa, Kaden+Lager & Hawkins\Brown
| Published 17 March 2016

High-rises offer new opportunities. The use of wooden frames and construction systems in tall buildings is rising rapidly. We take a closer look at four European examples where local conditions have put their stamp on the buildings.

城市木业进展

VARIOUS PARTS OF the world are erecting taller wooden buildings than could ever have been imagined a few decades ago. The tallest so far is the recently built Treet, a 45 metre, 14-storey block in Bergen. There are even plans to build wooden structures that fall into the category of skyscrapers, i.e. buildings over 100 metres high. In Canada a detailed investigation, based on research, has shown that it is both technically and economically viable to build 30-storey buildings around a wooden structural frame and that this brings major environmental benefits. The American architectural firm Skidmore, Owings & Merrill has conducted a similar study of a 42-storey building.

In Sweden, as in the rest of Europe, stone, brick and concrete have long been the dominant construction materials for apartment blocks. A rapidly growing population in the large cities and a threatening climate crisis are forcing us to think along new lines. Greater use of wood as a construction material for buildings is part of the solution. Supply is good, the material is green and it is easy to work with in a factory or on a building site. The fact that living in wooden homes is also pleasant and comfortable has been confirmed by Technische Universität Graz in Austria.

Wood is a light material in relation to its strength, it has good thermal properties and is durable as long as it is constructed and maintained properly. Growing trees absorb large quantities of carbon dioxide, which is then stored in the material. Wood has little environmental impact during manufacture, transport and installation compared with steel and concrete. With smart forest management, harvesting and processing, increased use of wood can play a major role in mitigating the climate crisis.

Thanks to technical advances, wood is in a position to seriously compete with concrete and steel as a structural material for multi-storey buildings. To highlight the trend, we are showing how in

高层建筑带来新的机会。在高层建筑中木结构与构造系统的应用正在急速上升。让我们深入了解一下欧洲的四个案例，它们已将当地特色烙印于这些建筑上。数十年前无法想象，如今全球范围内许多地方都在建造更高层的木建筑。目前为止，最高的木结构建筑已达到 45 米，是一幢在挪威卑尔根的 14 层建筑。甚至有一些用木结构建造摩天大楼的计划，使其达到 100 米之高。在加拿大，一个基于研究的调查显示，技术上和经济上来讲建造一个高 30 层的木结构建筑是可行的且非常有益于环境。美国建筑事务所 Skidmore, Owings & Merrill 公司也进行了一个相似的 42 层木结构建筑的研究。

和其他欧洲地区一样，在瑞典，石材、砖块及混凝土长期占据着公寓大楼的主要建筑材料地位。急剧增长的大城市人口以及严峻的全球气候问题迫使我们思考并寻找新的出路。更好地利用木材作为建筑材料也是解决方案之一。木材供给充裕，环保绿色，易于施工。奥地利 Technische Universität Graz 大学证实，居住在木建筑中会令人心情愉悦并感到舒适。

木材相对于它的力学性能来讲是十分轻盈的，它有着极好的保温性能，若在建造并维护得当的情况下，木材还有很好的耐久性。生长的树木吸收大量的二氧化碳然后将它们储存起来。相比于钢筋混凝土，木材在制造、运输以及安装上对环境的影响很小。若是加上合理的森林管理、砍伐及加工，大量的使用木材将对缓和气候问题非常有效。

由于现代技术的发展，在多层建筑的结构上，木材到了可以和钢筋混凝土激烈竞争的状态。为了强调这个趋势，我们将展示四个不同国家的四栋高层建筑是如何因地制宜的被建造。其中有三个建筑是独立的，还有一个建筑位于密集的城市环境中。在挪威 Bergen 的这幢高层建筑以胶合木作为结构以箱体作为单位。在芬兰 Jyväskylä 的郊外，正在建造以木制结构箱体的 6 层至 8 层建筑。伦敦的建筑案例十分有趣，独创性的旋转十字型布局并采用了钢筋与交错层积材（CLT）混合结构。柏林的案例则是结合了一个 7 层和一个 5 层的建筑，并建在了密度较高的老城区。

然而，木结构建筑要做大做高依然还存在很多挑战。规模大和层数高的木建筑因为材料轻便将会面临风荷载的考验。同时，公寓之间的隔音也是个更为复杂的问题。意外火灾以及火势蔓延的风险必须最小化，且结构必须能够承受荷载和变形，不管何种建筑材料，都应该做到这些。

斯堪的纳维亚半岛是一个布满了诸多古镇的半岛，且这些小镇上有着各种用途的木结构房屋。火灾可以在这种密集的地方掀起毁灭性的灾难。瑞典早年就实施了一个防火规定，并在 1874 年颁布了防火法令并贯彻于全国各地。国立建筑规则在上世纪 40 年代被推行。从原则上讲，早年建立超过两层高的木结构房屋是不可能的，但是在 1994 年环境发生了改变，由于欧盟推行了一个新条例，允许建造此类高层高密度的建筑，只要防火安全要求达标。

承载和防火方面的结构必须有合格的防火措施去保持预期的建筑相关的功能效果。木结构建筑可以通过使用一些合适材料作为覆层，诸如石膏板。在火灾中，厚度大的木材表面会被烧焦以减缓燃烧的进度。运用木材烧焦程度的知识，承重结构在火灾中将会保持其完整性。这个原理被应用于卑尔根的那所建筑上，胶合木结构在建筑外部可见。对于加拿大 30 层木结构的研究来说，这也不失为一种合理的替代选择。

four different countries four high-rises have been constructed based on the local conditions. Three of the buildings are freestanding and one is located in a dense urban environment. In Bergen the high-rise was built from box units within a structural glulam frame. Outside Jyväskylä in Finland, six to eight-storey buildings are being built using structural box units in wood. The block we focus on in London has been given an interesting form through the ingenious rotation of a cruciform layout, which was made possible through a hybrid design using steel and CLT. The building in Berlin comprises one seven- and one five-storey block and stands in a dense area of older housing.

THERE ARE SEVERAL CHALLENGES to going larger and higher with wooden buildings. Large and tall buildings made from wood are light and so wind loads are a greater problem than for buildings in heavier materials. For the same reason, sound insulation between apartments is a more complicated issue. The risk of accidental fires and the spread of fire has to be minimised, and the structural elements must be able to handle loads and deformations, although this is true whatever the construction material.

Scandinavia is dotted with quaint old towns that are a jumble of wooden buildings with various uses. A fire in such a dense district can have catastrophic consequences. Sweden was early to adopt local fire regulations and since 1874 there has been a fire statute covering the nation's towns and cities. National building regulations were introduced in the 1940s. In principle, it was previously impossible to build higher than two storeys with a wooden structural frame, but in 1994 the conditions changed as the standards were brought in line with EU regulations, which are function-based. It is now permitted to go high and dense in wood, as long as the fire safety requirements are met.

Load-bearing and fire division structures must have sufficient fire resistance to maintain their intended function for the time stated in the building relations. Wooden structures can be protected through cladding in appropriate materials, usually plasterboard. In a fire, thick timbers char on the surface, which slows the burning process. Applying knowledge about the charring rate, load-bearing structures can be sufficiently dimensioned to retain their integrity in a fire. This method was used for the glulam framework that can be seen on the outside of the building in Bergen. It was also one of the alternatives in the Canadian study into a 30-storey building.

An increasingly common fire safety method is to equip the buildings with sprinkler systems, with apartments getting domestic sprinklers and more vulnerable locations having automatic sprinkler systems. The use of sprinklers allows more unprotected wood to be used, for example on façades.

PRACTICALLY ALL LOW-RISE HOUSES in the USA have long been built using timber-frame systems based on 2 x 4 inch joists. These systems have been standardised, which makes things simpler at every stage from sawmill to construction site. The same systems have also long been employed for wooden

buildings over several floors.

In Sweden too, timber-frame systems have dominated low-rise production over the past 50 years, but the level of standardisation has not been as high. When, in the 1990s, we began building multi-storey apartment blocks with a wooden structural frame, we copied the American way of building. This delivers light, material-efficient and flexible structures, but cannot normally be used for taller buildings higher than four or five storeys. The five-storey section of the building in Berlin used this system, with joists of 45 x 180 millimetres.

The development of solid wood systems began in the 1990s in many countries, with Switzerland and Austria leading the way. The first of these comprised low-grade sawn boards nailed, and later pegged, together in parallel, known as 'Brettstapel'. The panels are used in walls that are insulated on the outside. As a floor structure, they can be combined with cast concrete to form a composite construction that works as well as reinforced concrete, but with the wood panel replacing the rebar. To get the wood and concrete to work together, sections are cut out of the wooden panel, or it may be furnished with various types of mechanical tie, or sometimes both. A variant where the boards are screwed together was used in the building in Berlin.

During the first half of the 1990s, Cross Laminated Timber made its first appearance. In CLT, boards are layered at right angles and glued together. As sheets and panels, the material opens up many architectural and structural opportunities. Used in floor structures and walls, it can handle heavy loads and has a stabilising effect. Detailing and joints are easy to execute. Box units made from CLT are light and stable to transport. Sound insulation is simplified since the

对于建筑应该使用洒水器系统的普遍共识正在急速增长，有的公寓使用家庭洒水器以及许多易受到火灾威胁的地点已经开始使用自动化的洒水器系统。洒水器的使用允许了更多未做防火保护的木材得到使用，比如在外立面上。

事实上大多美国的低层建筑使用 2x4 作为托梁的木结构已经有很长一段历史了。这种系统被标准化，使得建设过程更加简单。相同的系统也被应用于多层木结构建筑之上。

瑞典也是如此，木结构系统主导了低层建筑将近 50 年，但是这种规范等级并不是特别高。在 1990 年代，瑞典开始学习美国的木结构方式建造多层木结构公寓。这带来了轻便、高效以及灵活的结构，但是这种方式通常不能运用于高于 4 层或 5 层的建筑上。柏林的五层建筑案例应用了此系统，不过是基于 45×180mm 的规格梁之上。

上世纪 90 年代，在瑞士与奥地利的带领之下，实木结构在许多国家的发展十分迅速。所谓的"Brettstapel"方式就是将低等级的锯材用钉子固定连接并平行组装。嵌板放置在墙体内以隔绝外部。对于楼板结构，它们可与模铸混凝土结合形成复合结构，用木板替代钢筋，也能达到加固混凝土的效果。为了让木材与混凝土共同使用，建筑剖面会加木板或者装配各种机械连接，或者同时都使用。在柏林的案例中，该栋建筑就使用了将大量木板钉在一起的方法。

90 年代的前期，交错层积材 / 正交胶合木（CLT）出现了。它将每个板层按顺序直角相交并胶合于一起。CLT 的出现为建筑结构行业带来了新希望。在楼板和墙的使用中，CLT 能够承受大的荷载并具有良好的稳定性能。细化部分以及连接部分能够实施得非常方便。CLT 制作的箱体结构轻便稳定方便运输。在双层楼板以及墙面的条件下，隔音问题也被简化。

所有四种参考案例都使用了 CLT。近些年来，致力于发展 CLT 混合结构的工作已经展开。CLT 与钢筋混凝土的结合使用，大大增强了其荷载与稳定性。对于楼板来说，CLT 可以与混凝土结合使用如同"Brettstapel"方式一样。伦敦案例的骨架也是由钢筋与 CLT 复合而成。

目前大约百分之十的瑞典公寓建筑均为木结构。胶合木、CLT 与复合结构这些可与其他材料相结合的材料似乎是木结构建筑向更高更大迈进的基石。使用箱体结构 / 盒子单元，木结构发展可大幅度工业化，质量将得到保障，且成本也将得到减缩。环境问题来看，大城市房源不足以及土地不足的问题都会推动高层大型木建筑的迫切需求。我们瑞典应该紧追时代的步伐。

以下我们将列出四种不同风格的欧洲建筑案例。

Treet by Artec in Bergen, Norway

世界上最高的木结构建筑在挪威卑尔根（Bergen），高约 45 米，共 14 层，拥有 62 间公寓，并分割成三个相似的部分（需了解原建筑）。外部胶合木结构以及两个混凝土稳固楼面用来承受水平以及垂直荷载。

4 个以 CLT 作为骨架的预制房屋模块被放在混凝土平台上用来组成混凝土地下室的天花板。外部结构则是在地桩基础上建造了 5 层楼高。再加上连接上一层的加固结构，一个特殊的部分被用来当做"承重楼层"。混凝土板也被用在此。第五层和最高处的模板被这个骨架所支撑着，而不是被最下面的模板支撑。这种形式的模块和结构被不断复制并往上安装，它以限制模块的垂直荷载并控制火灾。这种钢筋嵌入式结构，须保持在火灾的 90 分钟内能够承受住荷载。为了计算最大的尺寸，推测假设胶合木每一分钟烧焦 0.7mm。第二承重结构比如走廊阳台，必须在 60 分钟的火灾中保持其完整性。以及电梯井道由经过防火漆处理的 CLT 建造，并且给建筑备洒水器系统。在结构的外部，玻璃安装在两个外立面上，山墙端则覆上耐候钢板。

Puukuokka by Oopeaa in Jyväskylä, Finland

在 Jyväskylä 外，Puukuokka 的一个郊区，有三栋公寓楼使用木头建造，层高 6 至 8 层不等。第一栋于 2014 年建造完成。预制的 CLT 材料是由 Stora Enso 公司提供，它推动了多层城市建筑概念系统的发展。

此项目有一个明确的目标那就是试验一种可支付得起并且生态的新型建筑方式。预制的模板在室内装配保证了安装的质

walls and floors form double structures.

All four of our reference buildings use CLT in one way or another. In recent years, work has begun on developing hybrid CLT systems. These are systems in which CLT works with other materials, such as concrete and steel, to take up loads and provide stability. For floor structures, CLT can be combined with concrete in the same way as Brettstapel. The carcass of the building in London is a hybrid of steel frames and CLT panels.

AROUND 10 PERCENT of apartment blocks in Sweden currently have a wooden structural frame. Glulam, CLT and hybrid constructions where wood works with other materials are seen as the solution for larger and taller buildings. With box units, the process can largely be industrialised, quality can be guaranteed and costs kept down. Environmental considerations, the prevailing housing shortage and the lack of land in the big cities are all driving the increased construction of large and tall apartment blocks with a wooden structural frame. It is important that we in Sweden make sure we are part of this trend.

Below we present four European buildings of different styles and designs.

Treet by Artec in Bergen, Norway

The world's tallest wooden building is located in Bergen and stands around 45 metres high. It has 14 floors and 62 apartments, and is split into three similar parts as it rises. An external glulam framework and two stabilising floors in concrete handle vertical and horizontal loads.

Four prefabricated housing modules with a CLT carcass were placed on top of each other on a concrete deck that forms the ceiling of a basement level in concrete. The external framework was built five storeys high on piled foundations. Within this, and tied to the upper reinforced part of the framework, a special module was fitted to form what is known as a 'power storey'. A concrete slab was then cast onto this. The fifth and top housing module is thus supported by the framework, rather than the underlying modules. This pattern is repeated with other modules and the framework that is mounted above. This limits the vertical stresses on the modules and the consequences of any fire.

The framework, with its inset steel gusset plates, must be able to retain its load-bearing capacity for 90 minutes in the event of a fire. To calculate effective dimensions, it is assumed that glulam chars at a rate of around 0.7 millimetres per minute. Secondary load-bearing structures, such as corridors and balconies, must retain sufficient integrity after 60 minutes. The stairs and lift shaft are made from CLT treated with a fire-retardant lacquer, and the building is fitted with sprinkler systems. On the outside of the framework, glass has been fitted on two façades, while the gable ends are clad in weathering steel plate.

Puukuokka by Oopeaa in Jyväskylä, Finland

Outside Jyväskylä in the suburb of Kuokkala, three apartment blocks are being built in wood, six to eight storeys high. The first was completed in 2014. Prefabricated modules in CLT were supplied by Stora Enso, which developed the system under its Urban MultiStorey concept.

An express aim of the project is to trial a new construction process for affordable and eco-efficient housing. Prefabricated modules manufactured indoors and fully prepared for installation at the construction site guarantee high quality, avoid weather issues and mean minimal build time on site. Each apartment comprises two modules. One contains a living room and bedroom, plus a balcony. The other houses the bathroom, kitchen and hall. Installations are accessible from the corridor which, like the stairwell and lift shaft, is made from CLT.

In the apartments, the surface of the wood on the ceiling has been left visible, and the floors are also wood. The walls are clad in plasterboard, partly so they can be painted and partly to meet fire safety standards. The CLT frame serves both as a load-bearing and stabilising element, as well as providing a vapour barrier and partial heat insulation. A layer of concrete, with integral underfloor heating, is cast on top of the supporting floor to improve impact sound insulation.

The buildings are fitted with domestic sprinklers. Facing the courtyard, the façade material is untreated larch. The majority of the balconies are on this side, with most inset and fitted with sliding doors. This opens up the apartments to the surrounding countryside. On the street side, the spruce façade is treated with a dark paint. The façades were delivered as prefabricated units and hung using concealed fittings.

The project has been carried out with the support of a special 'lease-to-own' funding system. The apartments are underwritten by a state-backed loan at a low interest rate and for 20 years the tenants pay this loan off through their rent. After that period they then own their apartment. In 2015, the buildings were awarded the Finlandia Prize for Architecture and the Finnish equivalent of the Swedish Timber Prize.

城市木业进展

量，避免了天气带来的影响和提供了最短的现场安装时间。每个公寓由两个模块组成，一个包含客厅卧室外加一阳台。另外一组包含洗手间厨房和过道。安装从走廊入手，楼梯井、电梯井都是由 CLT 制成。

在这些公寓中，天花板上木材表面是可见的，地板也是木制的。墙面上覆盖了石膏板，可以涂上防火材料以符合防火标准。CLT 框架主要起到承重和稳定两方面作用，也提供了防潮及隔热功能。因为有底层的地热，混凝土层必须浇筑在支撑面板的最上方以提高隔音性能。

这些建筑装配内部洒水器。面朝庭院的外立面材料使用了未经处理的落叶松。大多数阳台也都在一边，并配备了移门。这使得公寓与周围乡村环境融为一体。在街道的一边，云杉制成的外立面被喷上了深色涂料，这些立面均为提前预制并使用隐藏装置悬挂配送。

这个项目的落实来源于一个"lease-to-own"的专项基金支持。这些公寓被一个国立贷款机构以低利率认购并且租客可用 20 年来支付的租金还清贷款。在那之后他们就可以拥有这个公寓。2015 年，这些建筑得到了 Finlandia Prize 建筑奖项和 Swedish Timber 的奖项。

C13 by Kaden+Lager in Berlin, Germany

在柏林的 Prenzlauer Berg 地区，有一栋含有公寓、办公、托儿所、餐厅等综合功能的建筑于 2012-2013 年被建立起来。它由两幢大厦组成，前面一栋为 7 层建筑，后面一栋为 5 层建筑。他们均为木结构结合混凝土地下层并带有车库。这些建筑被老式房屋环绕着。由于防火原因，它的入口以及电梯建立在外部混凝土的阶梯与坡道上，并且与边上的一栋老建筑相连。这是一个关于新木结构建筑与旧小区结合的重要案例。

所有的承重结构都是木制的。柱子均为胶合木，除了车库以及地面层的一些柱子用了混凝土加固。前栋建筑的承重墙由 85-100mm 厚的 CLT 制成。后栋建筑的承重墙是提前预制的，它是由矿物绝热棉填充的木结构，厚 180 毫米。为了满足 REI90 防火规定，所有的墙和胶合柱都需要覆盖两个 18 毫米厚的石膏板。外部的墙体则需要 18 厘米厚的石膏板并填充矿物棉，厚度分别为 140 毫米和 40 毫米，外部还需要加上一个 15 毫米的底泥层。

瑞典已经了解了此类建筑表面的问题。如果处理不够得当，水会渗透到底泥层去并且对内部石膏板或者木结构造成破坏。

楼面结构由混凝土和木材复合而成。这种 Brettstapel 板有 140 毫米厚，再覆盖上。一个 120 毫米厚的混凝土层覆盖，这保证了足够的承载力。为了满足隔音要求，还要加上 30 毫米厚的隔音层，外加 74 毫米厚内含加热管道的混凝土层。下层的木料经过了透明的防火处理。结构的总厚只有 404 毫米，比起瑞典常用的木结构要薄。木材与混凝土两种材料的混合结构，造就了更加高效的木结构 / 楼板结构方案。

The Cube by Hawkins\Brown in London, UK

英国最高的木建筑坐落于伦敦 Hackney 区的 Wenlock 街。这所名叫 The Cube 的建筑，高 33 米，有 10 层楼。

这所建筑非常惊人，它的楼面是由十字形轻盈可拆除的臂型结构组成。旋转式的造型有助于保护邻里间的隐私且有不同方向的露台。这些公寓通过巨大的窗户可以享受到充足的阳光。

楼体与电梯包含在一个混凝土核心结构中。此外，这个建筑也是 CLT 混合型的结构。钢筋结构和 CLT 板相互依附并被非常稳定的连接固定着。这种建造方式适合高层建筑。构架既轻盈又稳定，还能提供相当高的承载力。此复合结构易于安装，相比于纯钢结构或者混凝土结构建筑，它的碳排放量非常低。

建筑外立面覆盖着西部红雪松，外加一个大型砖格栅延伸至两个立面。

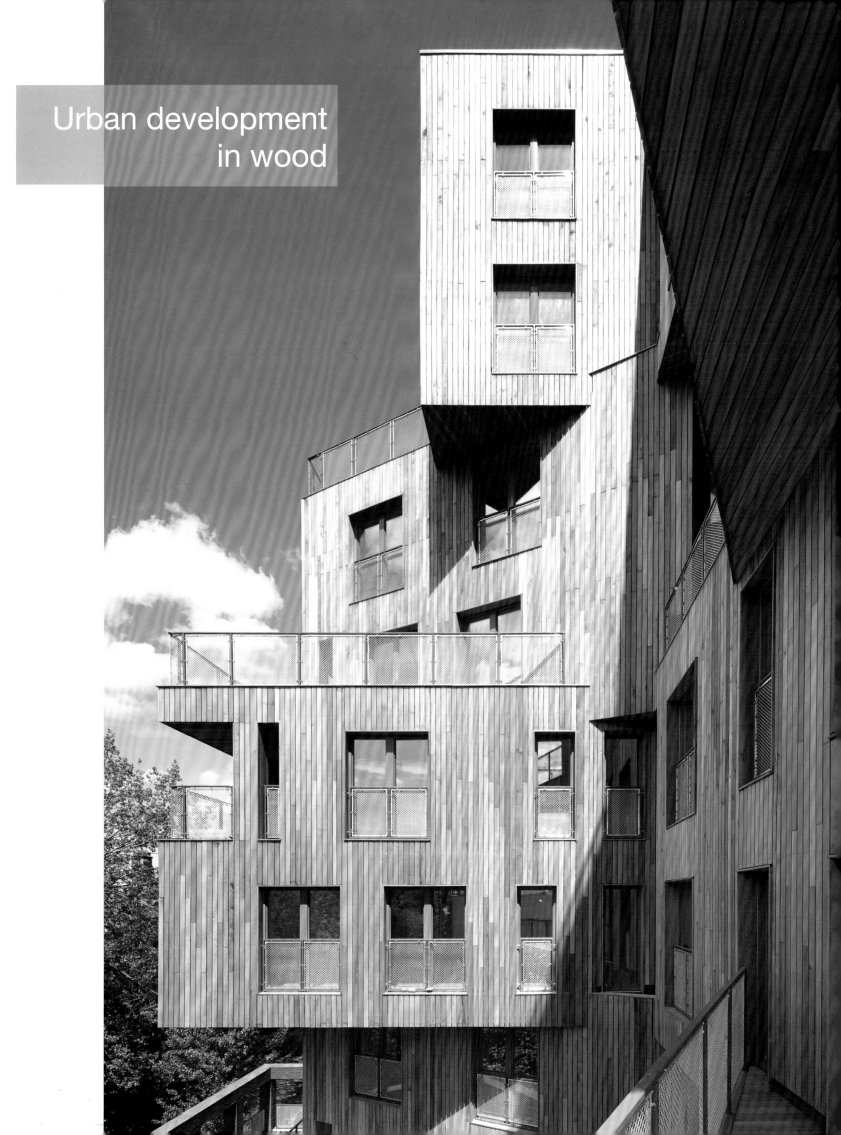

Urban development in wood

C13 by Kaden+Lager in Berlin, Germany

In Prenzlauer Berg, a district of Berlin, a building was constructed in 2012–13 that houses apartments, offices, a nursery, restaurants and more. It comprises two blocks, a front building with seven storeys and a rear one with five storeys. They are built in wood on a concrete basement level which includes a garage. The building is surrounded by older homes. For fire safety reasons, the entrances and lift are sited off external concrete stairs and ramps, and they are attached to an older neighbouring building. This is a prime example of how new wooden buildings can be inserted into older cityscapes.

Practically all the load-bearing structures are wood. All the columns are glulam, except for those in the garage and ground level, which are reinforced concrete. The load-bearing walls in the front block are CLT with a thickness of 85–100 millimetres. The walls of the rear block are a prefabricated, mineral wool-insulated, timber-frame structure, 180 millimetres thick. To meet fire safety regulation REI 90, both types of walls and the glulam columns are clad with two 18 millimetre thick plasterboard sheets. The external walls on both structures are 18 millimetre plasterboard clad in mineral wool, 140 and 40 millimetres thick respectively, with a 15 millimetre layer of render on the outside.

Sweden has seen problems with this type of façade solution. If it is done wrong, water can penetrate through imperfections in the render and cause damp damage to the underlying plasterboard or the wooden structural frame.

The floor structure is a composite of concrete and wood. The Brettstapel panel is 140 millimetres thick. On top of this, and tied to it, is a 120 millimetre concrete slab. This ensures sufficient load bearing capacity. To achieve the necessary impact sound insulation, there is then a 30 millimetre thick layer of acoustic insulation, followed by 74 millimetres of cast concrete, with inset heating pipes. The wooden underside is coated with a flame-resistant transparent fire safety treatment. The total thickness of the structure is only 404 mm, which is slim compared with the wooden floor structures that are usually preferred in Sweden. The hybrid structure of two materials, wood and concrete, creates an effective floor solution.

The Cube by Hawkins\Brown in London, UK

The UK's tallest wooden building stands on Wenlock Road in the London borough of Hackney. The building, called The Cube, is 33 metres high and has 10 storeys.

The architecture is striking. The floor plan is cruciform with slightly offset arms. Rotating them in pairs provides privacy from the neighbours while also creating terraces with a different orientation. The apartments enjoy plenty of natural light through their large windows.

Stairs and lifts have been incorporated within a concrete core. Otherwise, the building is a hybrid CLT construction. Steel frames enclose panels of CLT and these are joined with strong fixings. This way of building is suitable for tall buildings. The carcass is both light and stable, and provides a high load-bearing capacity. It is easy to erect, while at the same time delivering much lower carbon emissions than structures built entirely in steel or concrete.

The façade is clad in Western red cedar, with a large brick grid extending beyond two of the façades.

The Nominated for the 2016 Swedish Timber Prize

2016年瑞典木建筑奖提名项目

Text: Katarina Brandt | 校译：蒋音成 王恺
摄影：Åke E:son Lindman

| The 2016 Timber Prize in Sweden
| Published 25 November 2015

巨大的质量与技术优势体现在了 2016 年瑞典木建筑奖十个候选作品之中。

经过评委会的深思熟虑，在对 2015 年 1 月截止日期前提交的 139 个作品进行严格筛选之后，瑞典木业协会将于 2016 年 3 月 9 日公布第十二届瑞典木建筑大奖的获奖名单。这个竞赛对于采用何种结构形式保持开放的态度，只要木结构是其中主要组成部分即可，作品须已竣工且距今不超过四年，并可供评委会参观。

所有作品的评审标准为是否大体上符合一个好的建筑所具备的要求，关键点在于它的实际位置以及与周围环境的关系。作品所用的材料和细节也被考虑其中，包括各种功能如何被实现，当然，对于木材的使用也是必不可少的一点。

"随着建筑师和私人个体的关注，材料的质量以及建造的技术已经得到显著的进步。虽然建造业的发展变化进行得非常缓慢，但是自从这个奖项在 1967 年第一次设立以来，我们可以看到这种变化是巨大的。"瑞典木业协会的 Per Bergkvist 说，他从 1992 年后就负责此奖项。

大多数递交的参赛作品都涉及别墅或者度假屋，且这种现象在这个竞赛中始终存在着。

"我很惊讶的是我们收到如此少的关于公寓的作品，可能因为小规模的设计更容易获得一个好结果。显然我们需要更多地发展建造木制街区。" 评委会主席 Anders Svensson 说。

Anders Svensson 是一位拥有名叫 Caselab 公司的建筑师以及察尔姆斯理工大学的见习教授，他同时也是 2012 年颁布瑞典木建筑大奖时评委会的成员。

他说，"我很荣幸成为评委会的一员，在诸多作品中浏览筛选并一直全身心地投入来挑选其中的赢家真的是一个很有趣的过程。通过在建筑上，特别是在木结构建筑方面的讨论和交流经验的过程来看，我觉得这项工作非常值得。"

在评委会成员们独立浏览所有作品并对它们进行评分之后，有 31 个作品脱颖而出，评委们希望现场对这 31 个作品做出评审。

"在进行选择的过程中，我们做了一个集中的评估，筛选出了一些具有较高建筑价值并且能够代表当代木材使用情况的建筑。我们对建筑的造价也非常感兴趣，对每平方米的造价费用做出了比较。"Anders Svensson 说。

Per Bergkvist 和 Anders Svensson 都十分重视对这 31 个入围作品进行实地勘察这项漫长的任务。虽然从地理位置上来看这并不是很方便，但是评委会不会避开遥远的项目。特别是，几乎用上了所有能用的交通工具，在最高效的时间内穿越了瑞典的每一处。

"每一个我们考察的作品都令我们感到吃惊，它们表现出了木材使用的创新和令人兴奋的一面。大多数情况下，我们会去拜访建筑师和使用者，这是非常有价值的事情。最后的十强作品展现了现代瑞典木结构建筑进展的巨大优势。"Anders Svensson 表明。

自从这个奖项最初设立以来，很多不同种类的建筑都获得此殊荣。2012 年的获奖者为 Meter 建筑师事务所，他与委托方 Seitola-Gunnarssons 共同合作，打造了耶夫勒城外的 Tomtebo 森林桑拿项目。

在这之前的 2008 年，Brunnberg 和 Forshed Arkitektkontor 为 folkhem 公司设计的位于索伦蒂纳的 Östra Kvarnskogen 项目获得了该奖。作为此项目的主设计师，Kjell Forshed 相信木结构的发展有助于提高瑞典木建筑大奖的地位。

The Nominated for the 2016 Swedish Timber Prize

"我们非常荣幸获得瑞典木建筑奖。它证实了我们会向前一步并能够通过创新改革促使木结构建筑更好的发展。许多Östra Kvarnskogen 的建筑是十分传统的标准建筑。我们所做的就是让建筑通过细细的铁柱悬在空中,并且只有在建筑是木制的时候才能够实现。我们利用材料轻盈的特性创造了一些令人振奋的事。"

Kjell Forshed 希望 2016 年的奖项可以多关注一些较大规模的作品,而不是只是去奖励那种看上去十分引人注目的木建筑。

"我们需要额外推广木制品的日常生产,只有这样木制品才能更加受到重视而不仅仅是以其显著性和独特性存在。"

2016 年的奖项花落谁家将于 2016 年 3 月 9 日揭晓,届时第 12 届奖项的赢家将被授予 100000SEK 的奖金,所有提名的作品也将会被收录在《Architecture in wood – Träpriset 2016》这本书中并进行巡展。

The ten buildings shortlisted for The 2016 Timber Prize (Träpriset 2016) showcase strong advances in quality and technology.

SWEDISH WOOD WILL be announcing the 12th Träpriset architecture award on 9 March 2016, after the jury has considered the 139 entries that were submitted by the deadline in January 2015. The competition is open to any type of structure, as long as wood forms a significant part of the design. It has to be fully completed, no more than four years old and available for the Träpriset jury to visit.

The entries have been judged for the way they broadly meet the requirements of good architecture. The focus is on the actual site and the way the building relates to its surroundings and context. The Träpriset jury has also considered materials and details, and how various functions have been resolved. Plus the use of wood, of course!

"The quality of the materials and construction techniques has moved on quite considerably, as has interest among architects and private individuals. Changes are quite slow to occur in the construction industry, but looking back we can see that a great deal has happened since the award was first established in 1967," says Per Bergkvist of Swedish Wood, who has been responsible for Träpriset since 1992.

The majority of entries were submitted in the category villas and holiday homes – as has been the case throughout the history of Träpriset.

"It has actually surprised me that we've received so few entries in the apartment block category. It's perhaps still easier to achieve a good result on a smaller scale, and there may be a need for more development when it comes to building apartment blocks in wood," says the chairman of the jury, Anders Svensson.

Anders Svensson is an architect with his own company Caselab and Professor of the practice at Chalmers University of Technology. He was a jury member when Träpriset was last awarded in 2012.

"Sitting on the Träpriset jury is a real honour. Going through the submitted entries and being involved in the whole process all the way to choosing a winner is incredibly exciting. The work is also very rewarding in terms of the discussions and exchange of experience that take place on architecture in general and wooden architecture in particular."

ONCE THE JURY MEMBERS had independently reviewed all the entries and graded them, the list was honed down to 31 buildings that the Träpriset jury wished to inspect in situ.

"In the selection process, we made a joint assessment where we sifted out a broad range of buildings of high architectural merit that represented a contemporary use of wood. We were also interested in the money and made comparisons of the cost per square metre," says Anders Svensson.

Both Per Bergkvist and Anders Svensson highlight the often lengthy task of travelling around and visiting the 31 longlisted entries. The geographical spread was enormous, and the jury did not shy away from more remote projects. Practically every mode of transport was used to travel the length and breadth of Sweden in the shortest and logistically most efficient time.

"Every entry we examined up close surprised us, and showed new and exciting aspects of wood use. In most cases, we also got to meet the users and the architect, which was very valuable. The ten entries on the final shortlist reflect promising advances in modern Swedish wooden architecture," states Anders Svensson.

SINCE TRÄPRISET WAS first launched, a wide variety of buildings have won the award. The Tomtebo forest sauna outside Gävle won the award in 2012 and was designed by Meter Arkitektur in collaboration with their clients the Seitola-Gunnarssons.

过去的瑞典木建筑大奖的获得者：

2012 项目名称 – 设计者

Tomtebo forest sauna – Meter Arkitektur

2008 项目名称 – 设计者

Östra Kvarnskogen – Brunnberg & Forshed Arkitektkontor and Folkhem

2004 项目名称 – 设计者

Universeum – Wingårdh Arkitektkontor and Universeum

2000 项目名称 – 设计者

Trosa Skärgård holiday home – architects Natasha Racki and Håkan Widjedal

1996 项目名称 – 设计者

Zorn Textile Museum – architect Anders Landström

1992 项目名称 – 设计者

Information building for Vuollerim stone age village – architects Per Persson and Mats Winsa and curator Ulf Westfal

1988 项目名称 – 设计者

Villa Olby – architect Torsten Askergren and client Kerstin Olby

1976 获奖者

Kurt Tenning, civil engineer

1972 获奖者

Jan Gezelius, Architect SAR

1970 获奖者

Carl-Ivar Ringmar, Architect SAR

1967 获奖者

Carl Nyrén, Architect SAR

从 1988 年的瑞典木建筑奖开始，奖励规则发生了改变，这个奖被授予整个项目而不是一个人。

Before that, in 2008, Träpriset went to the housing development Östra Kvarnskogen in Sollentuna, designed by Brunnberg and Forshed Arkitektkontor on behalf of Folkhem. Kjell Forshed, chief architect for Östra Kvarnskogen, believes the increase in wood construction has helped to reinforce the status of Träpriset.

"We were highly honoured to receive the Träpriset award. It confirmed that we'd taken a step forward and introduced innovations that had resulted in better wooden architecture. The homes in Östra Kvarnskogen are actually quite traditional standard buildings. What we did was to have them hovering in the air on slim steel pillars, which was only possible because they were made from wood. We exploited the lightness of the material to create something exciting."

Kjell Forshed hopes that Träpriset 2016 will focus on a good example of something that works in larger-scale production, rather than rewarding a wooden building just for being spectacular.

"It's everyday production in wood that needs an extra push forward. That would be more significant than going for the remarkable and the unique."

THE WINNER OF TRÄPRISET 2016 will be presented on 9 March 2016, when the 12th award winner will accept the Träpriset statue and the prize of SEK 100,000. All the nominated entries will also be presented in the book 'Architecture in wood – Träpriset 2016' and in a touring exhibition.

The Nominated for the 2016 Swedish Timber Prize

PREVIOUS WINNERS OF THE SWEDISH TIMBER PRIZE – TRÄPRISET

2012
Tomtebo forest sauna – Meter Arkitektur

2008
Östra Kvarnskogen – Brunnberg & Forshed Arkitektkontor and Folkhem

2004
Universeum – Wingårdh Arkitektkontor and Universeum

2000
Trosa Skärgård holiday home – architects Natasha Racki and Håkan Widjedal

1996
Zorn Textile Museum – architect Anders Landström

1992
Information building for Vuollerim stone age village – architects Per Persson and Mats Winsa and curator Ulf Westfal

1988
Villa Olby – architect Torsten Askergren and client Kerstin Olby

1976
Kurt Tenning, civil engineer

1972
Jan Gezelius, Architect SAR

1970
Carl-Ivar Ringmar, Architect SAR

1967
Carl Nyrén, Architect SAR

From Träpriset 1988, the rules were changed so that the award would be given to a building rather than a person.

Raising the profile of the building's fifth facade
升华建筑物的第五立面

Text: Katarina Brandt | 校译：蒋音成　王恺
摄影：Michael Moran, Shigeru Ban architects, Aaron Hargreaves,
Rcbby Whitfoeld, Thomas Herrmann

| 英国、美国和德国的一些屋顶结构
| 福斯特事务所、坂茂、阿克尔曼和拉夫事务所。2014年12月1日发表
| Roof structures in United Kingdom,
　USA & Germany by Foster + partners, Shigeru Ban, Ackermann + Raff
| Published 1 December 2014

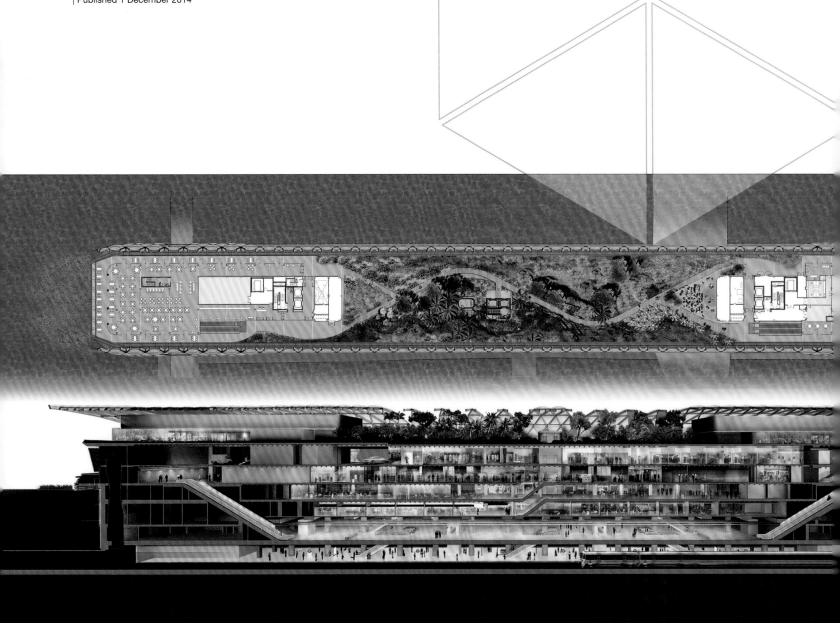

Advanced roof structures. The focus is now shifting up towards the building's fifth facade. Towards the roof and the opportunity to design creative structures with wood as a key component, not just structurally but aesthetically as well.

THE LIGHT WEIGHT. The formability. The suitability for large spans. There are many reasons for choosing wood when building a roof, according to Roberto Crocetti, Professor of Structural Engineering at the Faculty of Engineering, Lund University. He has also been involved in establishing construction company Limträteknik's new office in Malmö, which opened in September. Roberto Crocetti has noticed a growing interest in wood for large roof structures. This is true not least in countries where wood has historically not been as popular as in Sweden.

"Many architects realise the freedom that the material offers. Wood is relatively cheap. Another benefit in roof structures is that wood can stretch over large spans. I'd say it's one of the best materials in terms of its strength to weight ratio."

However Roberto also sees challenges. Wood is a living material that moves. As a structural engineer, it is crucial to predict the movements that might causes stresses. And this is particularly important when designing joints in wood.

"You need to have a feel for the material, to know how joints should be worked out and to protect the wood from the elements, either with various surface treatments or by enclosing the parts that are exposed to moisture. That way, the structure will last and last."

Raising the profile of the building's fifth facade
升华建筑物的第五立面

Another thing promoting increased use of wood is the interest in sustainability issues. This has long been a determining factor for Japanese architect Shigeru Ban. In April 2014, he was awarded the prestigious Pritzker Prize, often referred to as the Nobel Prize of the architectural world, whose purpose is "to honour a living architect or architects whose built work demonstrates a combination of those qualities of talent, vision and commitment, which have produced consistent and significant contributions to humanity and the built environment through the art of architecture."

ONE OF SHIGERU BAN'S latest projects is Aspen Art Museum in the US state of Colorado. The new museum, which opened in August, may not have the same spectacular exterior as many of Ban's previous international landmarks, such as the Centre Pompidou-Metz in France, with its billowing wooden roof. However, Aspen Art Museum remains a building that is in many ways typical of Shigeru Ban – not least when it comes to the roof structure.

The architect was initially unimpressed by the chosen location for the new museum building. Having climbed up onto the roof of a nearby building, he then realised that the site was not bad after all. From the roof of the building, he was able to look out across the rocky, tree-clad mountains and the many long pistes of the ski resort.

The experience led him to reassess the way visitors would approach the museum. Instead of placing the entrance on the lower floor and then continuing the movement upwards through the building, Ban did the reverse. The entrance is located up on the building's roof terrace and is reached via an external staircase or glass lift. From here you not only get a wonderful view, but also a glimpse of the impressive roof structure.

Shigeru Ban believes that the upward movement is particularly suitable for a place such as Aspen. In many ways, it calls to mind the sport of skiing, an activity that has made the resort famed the world over. Skiers take a lift up the slope, enjoying the view, before heading down one of the pistes.

Half of the roof terrace is covered with a glass roof, which is held up by a glulam structure. After this flourish, Shigeru Ban takes a more measured approach on the floors below, with the architecture being consciously restrained so as not to compete with the art on display.

As with many of his projects, he has chosen to work with the Swiss engineer Hermann Blumer, who specialises in complex wooden structures. The result is a fusion of architecture, technology and three-dimensional modelling.

The roof is made of glulam that is shaped into undulating waves. No metal ties have been used, due to the architect's wish to use renewable materials as far as possible. The Canadian company Spearhead Timberworks applied advanced CNC technology and huge precision to shape the curved glulam beams and ensure an exact fit, so that the elements locked into each other. The rolling waves of wood provide an exciting shadow play and depth on the terrace under the roof.

This type of advanced joinery technology is reminiscent of traditional Japanese wooden architecture. In theory the framework could be disassembled and reassembled, without damaging the materials.

IN DOCKLANDS, just outside the centre of London, Canary Wharf Crossrail Station sits like a ship in dock. The building is wrapped in a 310 metre glulam structure – a gigantic cover that makes up both the facade and the roof. The combined station building and retail centre was designed by Foster + Partners. The final parts are now being

人们的视觉焦点正望向先进的屋顶结构,即建筑的第五立面,望向使用木材设计充满创造力的屋顶的机会。作为建筑一个重要组成部分,屋顶不仅要满足结构需要,还美观。

"重量轻,可塑性强,适合大跨度结构,当我们建一个屋顶的时候,有很多理由使我们选择木材。"瑞典隆德大学工程学院结构工程专业的教授罗伯托·克罗切蒂说。他还参与了马尔默 Limträteknik 建筑公司的新办公楼的建设,该楼在九月份投入使用。罗伯托·克罗切蒂注意到,越来越多的人对于使用木材建造大型屋顶结构产生了兴趣。这个趋势不仅存在于瑞典这样有着使用木材的悠久传统的国家,也体现在其他的国家。

"许多建筑师意识到了木材提供的巨大创造空间。木材相对比较便宜。将木材运用于屋顶结构的另一个优点是,木材可以延展形成很大的跨度。我想说,从材料的强重比(强度与重量的比值)来看,木材是最好的材料之一。"

但是罗伯托也看到了挑战。木材是会产生位移的"有生命"的材料。这些位移可能产生应力,因此作为一名结构工程师,预测位移至关重要,尤其是设计木结构节点时。

"你需要对材料有一种感觉,来知道怎样解决节点问题,通过表面处理或包裹受潮部件来保护木材免受各种因素的影响,只有这样木结构才会经久耐用。"

另一个促进木材使用的原因,是木材有益于可持续发展。这对于日本的建筑师坂茂来说,一直是决定性的因素。2014 年 4 月份,他获得了久负盛名的普利茨克建筑奖奖,号称建筑界的诺贝尔奖。这个奖项是为了嘉奖在世的建筑师,其作品必须体现天赋、眼界、付出等品质,并通过建筑艺术始终为人类和建筑环境做出巨大贡献。

坂茂最新的项目是在美国的科罗拉多州的阿斯彭艺术博物馆。这个新博物馆在八月开放,它可能并不像坂茂之前的国际地标项目一样有着吸引眼球的外表,比如法国梅兹的蓬皮杜中心的波浪形木结构屋顶。然而,阿斯彭艺术博物馆仍是一栋典型的坂茂建筑——尤其是屋顶结构。

坂茂一开始并没有觉得这栋新博物馆的选址有什么特别。在爬上了旁边一栋建筑的屋顶后,他才意识到这个位置不错。从这栋楼的屋顶上,他可以看到布满岩石、树木葱茏的山脉和滑雪圣地的许多滑道。

completed ahead of the approaching inauguration in spring 2015. The building reflects the area's past as a centre for the UK's global merchant fleet. The site is now being integrated into London's cityscape and the growing commercial district of Docklands.

The initial work began back in May 2009 with construction of the dock that allowed the creation of the four storeys 28 metres below the water.

The glulam structure is protected by transparent air cushions in self-cleaning EFTE plastic. The cushions are fixed with aluminium strips to the underlying triangular beams. Despite the curved shape, the structure only makes use of four arched glulam beams. "The glulam frame provides a warm and natural counterpoint to the surrounding buildings in steel and glass," says Ben Scott, architect at Foster + Partners.

"Wood is natural, sustainable and welcoming. It's also relatively simple to shape the material to follow the roof's complex geometry."

On the top floor, closest to the roof, there will be a botanic garden. This is made possible thanks to the EFTE plastic in the roof, which creates the perfect microclimate for growing. The idea is that the garden will contain a selection of the tropical plants that once reached the UK via ships mooring up at Canary Wharf. "We've left the roof structure open in places, partly to let light in, but also for natural irrigation from the rain."

THE RIVER NECKAR that flows through Neckartailfingen, 25 kilometres south of Stuttgart, is a major influence on the German municipality. It has, for example, lent its name to the new Neckarallee Festival Hall which opened in 2013. Ackermann + Raff designed the building, whose wooden roof structure roots it firmly in its surroundings.

"The idea was to recreate the sense of walking along the nearby riverside promenade, Neckarallee," explains Julia Raff, architect at Ackermann + Raff.

"We were inspired by the pattern formed when branches and twigs overlap each other. Realising the idea in wood seemed like an obvious choice."

The roof stretches over the building in a diamond grid. Internally, white lacquered birch plywood fills in each diamond. The panels help improve the acoustics and have been placed at different angles to create life and movement. The surrounding areas of glass accentuate the roof's sense of lightness.

"The light roof structure has allowed us to keep the large assembly hall and the overhang outside the entrance free from supporting pillars," says Julia Raff. Externally, the roof is clad in fibre cement panels. All the visible wood is treated with a glaze. In the overhang outside the entrance, some of the diamonds have been left open to let in light from above.

这个经历让坂茂重新考虑游客进入博物馆的方式。他没有将入口布置在低层,逐渐向上行走穿越建筑,而是把入口设在了屋顶平台上,让人们通过一个室外楼梯和玻璃观光电梯抵达。在这儿你不仅能看到极好的风景,还能一瞥这令人印象深刻的屋顶结构。

坂茂相信向上的动线尤其适合用在像阿斯彭这种场所。它在各方面提醒人们,正是滑雪这项运动使这个度假胜地闻名世界。滑雪者乘坐电梯到达滑雪坡顶,在从滑道滑下来之前欣赏这风景。

一半的屋顶平台覆盖了玻璃屋顶,这些屋顶通过层板胶合木结构支撑。在这个大胆的动作之后,坂茂处理其他楼层更加低调,使建筑物的风头不至于盖过艺术展品。

坂茂在许多项目中都与擅长复杂木结构的瑞士工程师赫尔曼·布鲁默合作。最终的成果往往是建筑学、技术和三维模型的融合。

这座建筑的屋顶由波浪形层板胶合木构成,没有用一个金属连接件,因为建筑师想要尽量使用可再生的材料。加拿大先锋木构(Spearhead Timberworks)公司运用先进的数控机床精密制造出曲线形的胶合梁,保证精密无误,从而使这些构件相互无缝连接。波浪翻滚的木屋顶在下面的平台上投射出激动人心的阴影。

这种先进的节点连接技术,使人联想到传统的日本木结构建筑。理论上这种构架可以被拆卸和重装,不会损坏原材料。

在伦敦中心外的码头区,金丝雀码头车站像一艘船停在码头。这栋建筑被一个 310 米长的胶合木结构包裹形成了立面和屋顶。这个车站和零售商场综合体是福斯特事务所设计的,施工要赶在 2015 年春天的落成典礼之前完成。这栋建筑反映了这个地区曾经作为英国国际商船的中心,而现在这里已成为伦敦市景的一部分,也是码头区一个新兴的商业区。

项目始于 2009 年 5 月的码头建设——在水底建 28 米深的四层楼的设想。

自洁式 EFTE 膜中的透明气垫保护了层板胶合木结构。这些气垫通过铝条固定到三角梁上。这个曲面屋顶结构只用了 4 个拱形的层板胶合木梁。

"这个层板胶合木屋顶与周围的钢和玻璃建筑形成了温暖自然的反差,"福斯特建筑师事务所的建筑师本·斯科特说道。

"木材是天然的、可持续的和受欢迎的材料,也相对更容易做出复杂的屋顶形态。"

在顶层靠近屋顶的地方,会有一个花园。这得归功于屋顶上的 EFTE 塑料,它营造了适合植物生长的小气候。花园里将种上一些热带植物——很多年前这类植物曾通过金丝雀码头运到达英国。

"我们在屋顶结构的某些部位开了口子,一方面是想让阳光进来,另一方面是想让雨水浇灌。"

流经斯图加特南 25 公里内卡尔泰尔芬根的内卡河,对市政影响颇大。举个例子,2013 年开放的新的 Neckarallee 节庆大厅便借用了它的名字。阿克曼-拉夫事务所设计了这栋建筑,其木结构屋顶牢牢扎根在了周围环境当中。

"Neckarallee 的设计理念是为了营造一种在河边闲庭漫步的感觉。"阿克曼-拉夫事务所的建筑师茱莉娅·拉夫解释道。

"我们的灵感源于树枝交错的样式。所以采用木结构也似乎是顺理成章的。"

这座屋顶以菱形格栅的形式延伸至整个建筑。天花板上,烤漆的桦木胶合板填充了每一个菱形格子。胶合板不仅改善了声学效果,而且每一块的角度都不同,产生了活泼的动势。立面围护结构的玻璃衬托出了屋顶的轻盈。

"正是由于屋顶的轻盈,使大厅入口外的悬挑结构不需要柱子支撑。"茱莉娅·拉夫说道。

有可见的木材上施了一层釉料。在入口悬挑的部分,有一些菱形格子是镂空的,光线可以从上面透进来。

Raising the profile of the building's fifth facade

木材的优点

· 一种灵活的材料,有无尽的用法和用途
· 强韧、可塑性强,适用于大负荷、大跨度
· 稳定不变形,容易加工
· 经久耐用,耐风雨侵蚀和空气污染
· 天然环保,因为它是一种可再生和可再循环利用的原料。
· 耐火性能比其他许多材料都要好。
· 易于运输和安装。

木结构屋顶的三座新建筑

美国科罗拉多阿斯彭的阿斯彭艺术博物馆
建筑设计:坂茂
承建单位:特纳建筑公司
业主:阿斯彭艺术博物馆
造价:约 3 亿 4 千 4 百万瑞典克朗

英国伦敦金丝雀码头车站
建筑设计:福斯特事务所
承建单位:奥雅纳工程顾问公司
业主:金丝雀码头建设公司
造价:约 12 亿 5 千万瑞典克朗

德国内卡尔泰尔芬根 Neckarallee 节庆大厅
建筑设计:阿克曼+拉夫事务所
承建单位:Seyfried and Wiedemann, Frickenhausen and Wiehag GmbH
业主:Neckartailfingen 市政
造价:约 3 千万瑞典克朗

FÖRDELAR MED TRÄ

• A flexible material with countless possibilities and applications.
• Strong and formable. Copes with heavy loads and large spans.
• Dimensionally stable and easy to work with.
• Durable, even when exposed to weather, wind or air pollution.
• Eco-friendly, since it comprises a renewable and recyclable raw material.
• Resists fire better than many other materials.
• Easy to transport and fit.

THREE NEW BUILDINGS WITH WOODEN ROOF STRUCTURES

Aspen Art Museum in Aspen, Colorado, USA
Architects: Shigeru Ban Architects
Contractor: Turner Construction Company
Client: Aspen Art Museum
Cost: Approx. SEK 344 million

Canary Wharf Crossrail Station in London, United Kingdom
Architects: Foster + Partners
Contractor: Arup
Client: Canary Wharf Contractors Ltd
Cost: Approx. SEK 1,25 billion

Festivalhall Neckarallee in Neckartailfingen, Germany
Architects: Ackermann + Raff
Contractor: Seyfried and Wiedemann, Frickenhausen and Wiehag GmbH
Client: Gemeinde Neckartailfingen
Cost: Approx. SEK 30 million

choose wood where it is exposed to our human senses.

The planned extension to the Stockholm art gallery Liljevalchs Konsthall may be a work by the internationally renowned architect Kengo Kuma. His firm Kengo Kuma & Associates is involved in one of the teams competing for the job.

When do you choose to use wood?

"We choose wood whenever the material is available and we like to use it in spaces where the surfaces are solid and where they are exposed to our human senses."

What is the difference between European and Japanese wooden architecture?

"The biggest difference comes from the size of the trees. It used to be that only small trees were available in Japan. This has resulted in a refined joinery technique developing in the country."

Will wood be used for the Liljevalchs extension, if you win the competition?

"We've just started researching and analysing the project, so no decisions have been made as yet. But it would be great to use local wood."

What is your view on the increasing move towards prefabrication in the wood manufacturing industry?

"I believe fully exploring the technique could open up a host of new opportunities. There has long been a tradition in Japan of ensuring that wooden structures can be disassembled and reassembled without damaging the material. This has been achieved using complex joinery techniques that require no mechanical fixings. If prefabrication can learn from such ancient wisdom, the future could be very interesting indeed."

我在人们可以感知到的地方使用木头

国际著名建筑师隈研吾的事务所参与了斯德哥尔摩 Liljevalchs 艺术馆扩建项目的投标，这个项目可能成为他的又一作品。

——您为什么选择运用木材来设计？

"只要木材适用，只要空间表面是实的，只要在人们能够感知到的地方，我就会选用木材。"

——欧洲和日本的木建筑有何差异？

"最大的差别在于树木的尺寸。过去日本只有较小尺寸的树木，这客观上促成了精细化木工技艺在日本的发展。"

——如果您赢得 Liljevalchs 艺术馆扩建项目的投标，会使用木材来设计吗？

"我们才刚开始接触和分析这个项目，还没做出任何决定。但如果能运用瑞典本地的木材，无疑是极好的。"

——您对日益发展的木结构建筑预制工业化怎么看？

"我深信，预制装配技术的发展将创造出更多的机会来。木结构只要材料没有损坏，就能拆卸和重新组装，这在日本有着悠久的传统。我们运用的是复杂的连接技术，完全不需要机械固定节点。如果现代预制装配能够从古人的智慧中汲取营养，前景必将一片光明。"

访谈 1：隈研吾

校译：蒋音成

Published 30 May 2013
Kengo Kuma
Kengo Kuma & Associates
隈研吾及其工作室
发表于 2013 年 5 月 30 日

访谈 2：王兴田

王兴田，1961年出生，日兴设计·上海兴田建筑工程设计事务所总经理、总建筑师，日本北九州大学博士，国家一级注册建筑师。

Designer's profile: Wang Xingtian, born in 1961, General Manager and Chief Architects of NIKKO, doctorate of Kyushu University, national first-class registered architect.

作者：蒋音成
校译：韩佳纹

1月22日的上海，阴冷多风。上海交通大学刘杰教授和欧洲木业协会项目经理蒋音成先生来到了上海兴田建筑工程设计事务所。步入大堂，在紧凑的空间中，精致的陈设，忙碌的设计师……一切都显得气韵十足又有条不紊，为这寒冷的季节平添了几份暖意。在入口处众多的日兴项目展示中，他们一眼便被深圳隐秀山居酒店项目的木构雨棚模型所吸引。（附项目图）

刘：您当时为什么会采用现代木结构来完成这个酒店入口呢？

王：我对木构建筑的钟爱好像是与生俱来的，而对其生态性是逐渐认识的。我认为建筑的设计价值，就是用中低成本的材料和适宜技术的建造，把环境、空间、结构等完美地建构组合，提升整个建筑空间的价值。木材就是这样一种亲切、生态的材料。对于有着木构建筑悠久历史的我们而言，现代木构建筑还属于一种重新认知的建构体系，我想在更多的设计中去尝试。

It was cold and windy in Shanghai January 22. Prof. Liu Jie of Shanghai Jiao Tong University and Project Manager of European Wood paid a visit to NIKKO. The moment they walked into the lobby, they were welcome by the exquisite furnishings and bustling designers in such a compact space. It seemed everything was in a lively and orderly spirit, which brought in a sense of warmth. Among a number of NIKKO's project display panels at the entrance, they were caught by the wooden awning model of the Castle Hotel-Genzon in Shenzhen.

Prof. Liu: why did you apply modern timber structure to the hotel entrance?

Mr. Wang: I have an innate obsession with wooden architecture, while I've got to know the ecological features gradually. I think the architectural design value lies in the construction with mid-low cost materials and appropriate techniques to perfectly composite the environment, space and structure and to add values to the architectural space. Wood is such a intimate, organic material. Modern timber structure is a new construction system for us who have a long history of wooden architecture to perceive and I want to try in more design practices.

刘：我也是从 2004 年开始接触现代木构建筑的理念，我认为，中国的木构建筑要振兴，走传统的老路是行不通的，必须用新的技术与时俱进。

王：现在中国社会对现代木构建筑仍然存在较大的误解，很多人认为做木建筑会导致森林材资源的枯竭。实际上，有计划有组织地管理和利用，比森林无人看管、自生自灭的状态，更具有生态的可持续性。此外，钢筋混凝土建筑材料在生产过程中消耗大量的能源和资源，同时也造成环境的巨大负荷。混凝土建筑拆除后，钢筋混凝土的不可降解性，在全生命周期中，对环境造成的负担更大。所以，我们重新认识木构建筑的生态意义是很有必要的，它不仅有利于环境的可持续发展，也将促进建筑业的多元化发展。我在多个项目中都尝试运用木构建筑，当然，并不意味着把传统的木构建筑搬过来，而是在当代语境下，同当代木构技术的有机结合与创新。目前在中国，熟悉、能够较全面地掌握木构建筑或从事木构建筑的专业人员并不多，所以现代木结构建筑要在中国推广发展，仍然有很长的路要走。

Prof. Liu: I've started to know modern timber structure since 2004. From my point of view, it doesn't work if we follow the traditional path to thrive the wooden architecture in China. We have to use new techniques and keep updated.

Mr. Wang: there still are misunderstandings to modern timber structure in China. Many people think the application of timber structure will cause deforestation. In fact, well-planned forest management and utilization is more ecologically sustainable than the unattended. Besides, tremendous energy and recourses are consumed in the production of steel and concrete, bringing huge burden to the environment. Therefore, it is necessary for us to recognize the ecological significance of timber structure, which is not only good for the environmental sustainability but also helpful for the diversity of the building industry. I tried to apply modern structure in many projects. But of course, that doesn't mean to copy the traditional wooden architecture, but to organically integrate the modern timber structure technology and innovation in the contemporary architectural context. Currently in China, there are not many professionals who have grasped the modern timber structure expertise or have been working on modern timber structure. The development in China is a long way to go.

刘：您在应用木结构的过程中，一直都追寻建筑的地域性、本土性。您是如何尊重城市文脉和地域环境、实现传统建构与现代建筑的契合？

王：地域建筑产生的机制、背景，比它的形式重要得多。为什么会形成不同的建筑形式，是基于当时的物质和技术条件。生活方式和社会背景等差异产生了特定的建筑形态。例如，外土墙内木构的土楼建筑是在客家人迁徙和躲避战乱的恶劣环境中产生的，当时最重要的是安全防御，需要同族的人们团结一致，抵御外侵，便形成了向心的、内聚的、血脉生活的封闭土楼，显然，这放在注重居住生活舒适性的今天自然是不合时宜的了。只有意识到形态之外的生成机制，才能理解当今的地域建筑真实诉求，现代的地域建筑与当下社会的物质基础、生活方式和技术水平是紧密关联在一起的。

刘：这也是我为什么选择向当代建筑靠齐，而非历史建筑靠齐。我个人反思，现在的历史建筑保护在某种程度上是一个伪命题。中国在用现代的方式造历史的建筑，而原真性是很难的，时代是与时俱进的。

王：首先，既存的大部分传统建筑在居住环境、舒适度方面与当代生活的要求相去甚远，如果一味还原传统建筑形式，而不改善它的生存环境强行让人生活在其中，就失去了建筑是以人为本的真实内涵。还原传统建筑是还原记忆，还是重新使用？得到保护的传统建筑，如果缺少了其中人的生活也就失去了活力。另外，保护要分级分类，哪些应该作为文物来重点保护，哪些可以进行一些修缮和改造能够继续被现代人使用。总之，我们要把过去和现在关联起来，对历史的尊重不等于完全跪拜，而是把这种尊重举起来，形成历史和现代的对话。同样我们虽然有着历史悠久而独树一帜的木构建筑传承和技艺，而对当代木构建筑的技术现实，应放下沉重的传统思维，积极借鉴创新，这可能才是我的设计出发点。

the building industry. I tried to apply modern structure in many projects. But of course, that doesn't mean to copy the traditional wooden architecture, but to organically integrate the modern timber structure technology and innovation in the contemporary architectural context. Currently in China, there are not many professionals who have grasped the modern timber structure expertise or have been working on modern timber structure. The development in China is a long way to go.

Prof. Liu: As I know you've been perusing the architectural rationality and nativeness. How do you respect the urban context and regional environment to reach the conformance between traditional construction and modern architecture?

Mr. Wang: The mechanism and background of regional architecture are more important than its pattern. Various architectural forms are based on the substantial and technical conditions. Life style and social background decide a certain architectural pattern. For example, Hakka earth building with rammed earth exterior walls and wooden interior structure emerged when Hakka people immigrated to get away from wars. Defense was of the most importance which required unity of the clan so that it generated the concentric, centripetal and enclosed Hakka earth and its clan life. Obviously, it is not suitable for people's life today that demands comfort most. As long as we realize the mechanism behind the form and pattern, we may have a chance to understand the real request from regional architecture. Modern regional architecture is closely associated with the substantial basis, the life style and the technology of the current society.

Prof. Liu: this also is why I choose to keep up with contemporary rather than historic architecture. In my reflection, the historic building protection is to some extend a pseudo-proposition. It is difficult to keep the authenticity to renovate historic buildings in a modern method because the time changes.

Mr. Wang: First, most existing traditional buildings can hardly meet the contemporary life requirements in living environment and comfort. If we blindly restore the traditional architectural forms and neglect the living environment, we will lose the people-centered content. Is the goal of restoration of the traditional buildings to restore the memories or to reuse the buildings? A protected building will lose its vitality if no occupants live in it. Besides, I suggest classifying and categorizing the building protection work, to clarify which types of building should be protected as historic relics and which can be reused after rehabilitation and renovation. In a word, we should connect the past with the present. To respect the history doesn't mean to get on our knees but to lift the respect and initiate a dialogue between the past and the present. Likewise, although we have a long and unique history of wooden architecture and techniques, we should get rid of the outdated thinking and embrace the innovation, which is the starting point of my design.

佛教建筑是中国木结构运用非常重要的一个领域，唐宋以降多个时期都留下了许多经典的木构建筑文化遗产。近代以来，中国传统木结构建筑的发展受到能够塑造崭新空间的新材料和新结构的冲击。欧美等发达国家逐渐兴起的以胶合木为代表的工程木材，以其优良的防腐、防潮、一定的阻燃性和稳定的力学性能成为了木构建筑的新型材料。

2009年，杭州市运河集团在承建香积寺工程中，上海交大安地建筑设计有限责任公司刘杰工作室有幸设计中标，在古运河旁边的大兜路历史街区，着意恢复古代伽蓝七堂制的佛教寺院空间的同时，主体殿堂运用胶合木结构设计营建了杭州新香积寺。

胶合木在当代佛教建筑中的设计应用
——杭州香积寺复建规划设计

The Design Application of Glued Laminated Timber in Contemporary Buddhist Architecture
Hangzhou Xiangji Temple Rehabilitation Planning Design

作者：任琳哲 刘杰 | 校译：任琳哲 东鸿

Buddhist architecture is a very important wood using field in China, Tang and Song Dynasty has left many classic wooden architectural culture heritages. Modern times, the development of the Chinese traditional wooden building is impact of the new material and new structure which is able to shape a new space. Glulam, emerginged in Europe, the United States and other developed countries, with the excellent anti-corrosion, moisture-proof, flame retardancy and stable mechanical performance, has become a new material of wooden architecture.

In 2009, Liu jie Architectural Design Studio of Shanghai Jiaotong University Andi Architectural design co., LTD. designed the Xiangji Temple, which is next to the Beijing-Hangzhou Canal, the main halls were constructed by glulam and restored the ancient buddhist temple spacethe.

寺庙俯视

图 2-a 天王殿入口

一、香积寺的历史背景

香积寺位于杭州市拱墅区香积寺巷 48 号，是拱墅区历史上著名的寺庙，还是京杭大运河杭州起始端的第一座佛寺，又被称做"运河第一香"（图 1-a、b）。北宋太平兴国三年（978 年），柯氏舍宅为寺。元朝末年，为大火所毁，后虽几经重建，但最终只留下石塔茕茕孑立。现存的一座石塔为清康熙五十二年（1713 年）所建。

二、复建的规划设计理念

复建香积寺的基地位于香积寺原址以西，京杭大运河以东区域，基地面积 16855 ㎡，其中建筑基地面积 10971 ㎡，寺庙建筑部分由 16 栋 1~3 层殿堂组成，总建筑面积 13681 ㎡。其中地上建筑面积 10137 ㎡，地下停车场面积 3544 ㎡。（图 2-a、b；图 3）

1. 复建理念

历史上的香积寺几经兴废，但却并未留下太多的历史信息。今天复建香积寺，在定位在当年香积寺基本时代特征的前提下，而予以适当的创新，使其具备今天中国建筑文化的印记。

2. 材料与形制

建筑的结构部分采用胶合木，而外部则通过草架营造出传统屋面的举折生起，使得现代空间和传统风貌有机融合（图 4）。在室内，人们感受到的是开敞的室内空间和工程木的表面质感

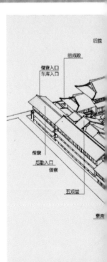

图 3 鸟瞰图

图1-a 香积寺地理位置

图1-b 香积寺新、旧址

1. Historical background of Xiangji Temple

Xiangji Temple locates in NO.48, Xiangji lane, Hangzhou Gongshu District, it was a famous temple in history, the first temple of the Beijing-Hangzhou grand Canal, named "The First Incense"(figure 1-a,b). It was donated by the Ke family in Northern Song Dynasty (in 978) and destroyed by fire in Yuan Dynasty. There remained only a stone tower reconstructed by Qing Dynasty (1713).

2. Design concept of reconstruction

The temple area is 16855m², the construction base area is 10971m², the total construction area is 13681m², the above ground floor area is 10137m², the underground area (parking lot) is 3544m².(figure 2-a,b;, 3)The temple consists of 16 layer 1 ~ 3 halls.

1) Design concept

Xiangji Temple has been rised and declined in history, did not leave much historical information. Today, we rehabilitate and innovate it, premising in the basic era characteristics, making it have the stamp of the Chinese architectural culture today.

2) Material and shape

Glulam structure inside, traditional structure external, the halls present a traditional roofing with folding arise, making the modern space and the traditional features integrate organic.(figure 4) Indoor, we can feel the open interior space and the unique flavor

图2-b 天王殿入口

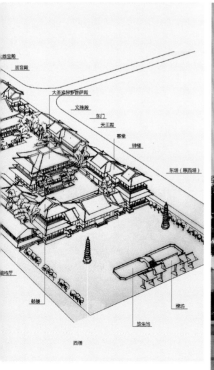

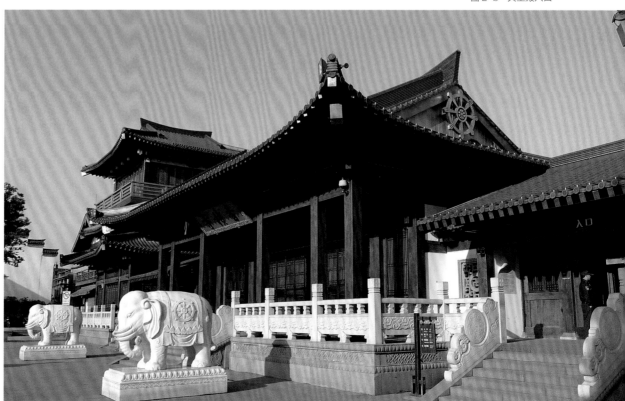

图4 天王殿翼角

所形成的独特韵味（图5）；在室外，传统的外观则与杭州这座古城及千年运河交相辉映，延续着杭州香积寺新的历史。

3. 功能

胶合木材料强度较高，可以获得更大的跨度，为礼佛朝拜提供了更多的空间自由度；其结构性能稳定，防腐、防火等性能较好，能使结构更加安全耐久。胶合木材质的表面质感相比传统木材要好，建筑建成以后格调高雅，更是成了运河边上一道亮丽的风景，承担起旅游和城市广场的作用。

4. 布局

香积寺是在北宋年间，由柯氏舍宅而成。因此，寺庙的最初布局较为简单，只是一般宅第庭院形式。到了南宋，禅宗的发展在江南达到了历史上的一个巅峰，宋宁宗仿效印度"五山五精舍"钦定了江南禅寺的等级，设禅院"五山十刹"制度。因此，香积寺的殿堂及附属建筑的配置基本沿袭宋代形成的"伽蓝七堂"制度。复建后的香积寺中轴线上，从最南端的石牌坊算起，依次有放生池（石桥）、寺前广场（东、西石塔）、天王殿、一进庭院广场、大圣紧那罗王菩萨殿、二进庭院广场、大雄宝殿、三进庭院广场、藏经阁（含法堂、方丈院）、后花园等建筑与庭院空间。（图6-a、b）

formed by surface texture of glulam;(figure 5) Outside, traditional appearance and the ancient city alternates with the Beijing-Hangzhou grand Canal, lasting a new history of the new Xingji Temple.

3) Function

The material strength of glulam is quite high, it can obtain a larger span than traditional wood, providing more space for worship; Its structure performance is stability, and it has a nice anticorrosion and fire prevention performance, making the structure more safety and durability. The elegant surface texture make the Xingji temple become a beautiful scenery of the riverside, It take on the role of tourism and city square.

4) Layout

Xiangji temple was donated in the Northern Song Dynasty. The layout was very simple at that time, just a courtyard. In the Southern Song Dynasty, the development of Zen reached a peak at south of the Yangtze, Emperor Ningzong of Song followed "Five Mountain Five Kitagiri" from India, (1) worked out the level of temples, therefore, the configuration of Xiangji temple formed this level of Song Dynasty (3).From south to north, there are Stone Archway, Free life pond, the temple square, the Heaven KingHall, the frist courtyard square, the Bodhisattva Hall, the second courtyard square, the Daxiong Hall, the third courtyard square, the Tripitaka Sutra Pavilion and the back garden.(figure 6-a,b)

图6-a 东立面图

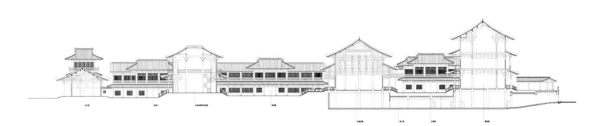

图6-b 整体剖面图

图 6　大雄宝殿室内

图 7 大雄宝殿室内

三、建筑设计

1. 建筑形式

中轴线上所有殿堂除大圣紧那罗王菩萨殿外都使用胶合木结构，整个框架体系遵循中国传统木造的结构逻辑。人处殿内，既有对传统空间的感受，又因新的结构使得室内基本无柱，比传统寺庙殿堂空间更加宽阔，佛像设置有更大的自由度，香客礼佛空间增加；同时因为室内构件少，且在殿堂两个山面上开窗，增加了采光，空间的体验效果比传统殿堂更好（图7）。室外则延续传统木构造副阶所形成的回廊形式，建筑外部造型则通过草架取得传统屋顶的效果。

2. 新材料的运用

除大圣紧那罗王菩萨殿、钟鼓楼，其他的殿堂采用木材与混凝土的组合。在材料的质感上，木材偏向于柔软，有亲和力，混凝土则给人以比较坚实冰冷的感觉，它们的关系是相互补充，对立统一的。

3. 新的构造方式

简化传统榫卯节点 "斗栱是榫卯结合的一种标准构件，在古代的建筑者心目中，斗栱大概是被认为最为成功、最得意之作。"但为了保证结构强度，难免复杂而臃肿，难于达到现代空灵轻盈的造型要求。香积寺延续了传统建筑的体形、基座、坡顶，但对传统的斗栱形式做了抽象和简化（图8）。以木枋纵横叠置的构造出现，继承斗栱的结构作用和力学原理，既满足了结构受力和施工的要求，又保留了传统节点的意向，唤起了人们对历史的记忆。（图9-a、b、c；图10-a、b、c）

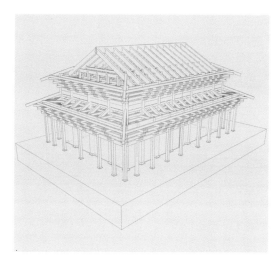

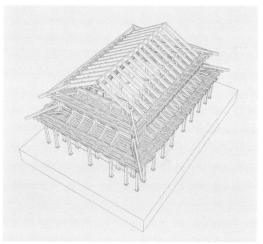

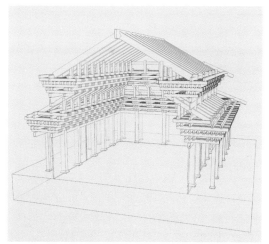

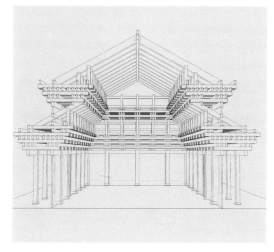

3. The architectural design

1) Architectural form

All of the halls used glulam structure in addition to The Bodhisattva Hall, the whole framework system follows the traditional Chinese wooden structure. People in the halls can get the traditional space feeling, and more pilgrimage space at the same time; the experience of space is better than traditional halls because of the less indoor component and skylights.(figure 7)

2) The use of new materials

All the halls are combination of glulam and concrete. On the sense of material, glulam soft and affinity, concrete express the sense with solid cold, both are necessary and should complement each other.

3) A new constructing mode

Simplify the traditional mortise and tenon joint

Tou-kung is a standard component of mortise and tenon joint, but in order to ensure the structural strength, it is complex and bloated, could not meet the requirements of modern airy lightness requirements. Xiangji temple lasts the traditional architecture of the body, the base and the top, but abstract and simplified Tou-kung.(figure 8) Glulam superimposed vertically and horizontally, not only inheriting the function and mechanics principle, meeting the structure and construction requirements, but also retaining the traditional node intention, reviving memories of the history as well. (figure 9-a,b,c; 10-a,b,c)

图8 大雄宝殿模型

图 9-a 大雄宝殿斗栱

图 9-b 大雄宝殿斗栱

图 9-c　大雄宝殿斗栱

图 10-a 大雄宝殿屋檐

图 10-b 大雄宝殿翼角

金属节点的借鉴与引入 胶合木的结构设计借鉴的是钢结构的设计原理，节点处理也多使用金属构造。钉或螺栓布置方便，受力明确，适用于受力较小的节点，这在寺内走廊或殿堂回廊随处可见。殿堂的梁柱结构，则是在胶合木节点处插入钢制连接件，并用螺栓与木构件连接。钢制连接件强度高，造型灵活，丰富了木构建筑的结构形式。（图11、12）

材料转换的处理 混凝土优越的防水、防潮性能和抗压强度，非常适合作为木结构建筑三段式的底部一段出现。胶合木与混凝土基础接触时，也要防止水或其他原因引起的潮湿对木构件的侵害。香积寺的殿堂在混凝土基础上预埋钢连接件，钢材的性能优越，强度高、刚度大，能够保证木构件与混凝土的可靠连接，实现材料转换并传递荷载，处理好防水防潮等问题，有效发挥木材和金属各自的优势。（图13）

斜梁 香积寺大殿的木框架结构继承了抬梁式的形式和意义，采用早期的大斜梁（后缩小成大叉手）形制表现得更为简洁，符合现代审美情趣，在简化斗栱的基础上采用大跨度的斜梁，进一步提升室内空间的品质。（图14、15）

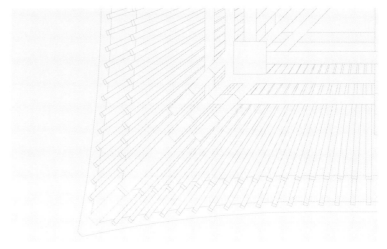

图10-c　大雄宝殿二层翼角方椽平面布置图

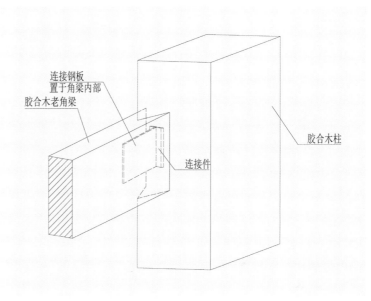

图11　柱与老角梁连接图

The introduction and reference of metal node
Gulam structure design principle reference of steel structure, the node processing also as metal construction. Screw's or bolt's arrangement is convenient and stress clearly, apply to force small node, this can be seen everywhere around the corridor. The beam structure of the halls is inserted with steel fittings in the node, bolt connection with wooden parts. Steel fittings have high strength and flexible modelling, enriching the structure of wooden buildings.(figure 11, 12)

The processing of material transformation
Concrete is very suitable for the bottom of a three-part wood construction for its superior waterproof, moistureproof performance and compressive streng. We should prevent water or others cause damp to the wooden parts when glulam contact with the concrete foundation. We embedded steel fittings in concrete, its superior performance, high strength and stiffness can ensure the reliable connection of the wooden piece and concrete, to achieve material transformation and loads transfer, dealling with the problem such as waterproof, effective play the advantages of wood and metal respectively.(figure 13)

Oblique beam
The wood frame structure inherits the traditional architecture's form and meaning, the large oblique beam (then changed into big hand fork)in the early is concise, in line with the modern aesthetic temperament and interest, the simplified Tou-kung and the large span enhance the interior space quality further.(figure 14, 15)

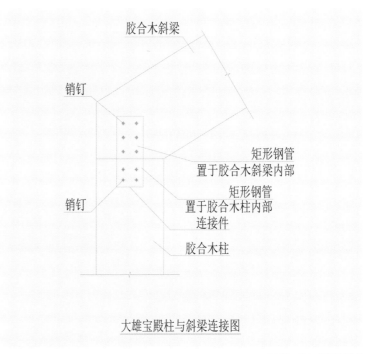

大雄宝殿柱与斜梁连接图

图12　柱与斜梁连接图

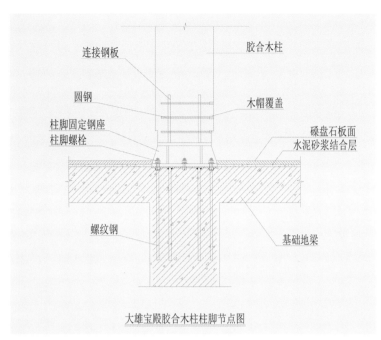

图 13 胶合木柱柱脚节点图

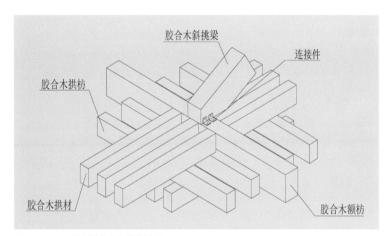

图 15 天王殿柱斜挑梁与额枋连接图

图 14 天王殿斜梁

图18 檐口

草架 是指安于平闇、平棊、藻井之上，未经细加工的，起到负荷屋盖重量的梁栿。在中国古建筑中，此类构件隐蔽在天花板之上。香积寺的四座大殿都运用了草架（图16、17），其隐蔽在斜梁及天花之上，塑造出传统的凹曲大屋顶。

减少环境对木材的破坏 外界环境会对木材造成影响，特别是雨水。虽然胶合木经过了防腐处理，但其断面不具有此性能。因此，在檩条断面处使用了铜包覆，既减少了雨水对木材的破坏，又呈现出耐人寻味的细部设计。（图18）

结 语

胶合木作为一种既有原木的特性（色、香、材质感等），又具有稳定的力学特性的材料，为宗教建筑营造超脱于传统形式的空间和造型，使人更加深刻地感受到宗教建筑的魅力。香积寺的设计是对胶合木结构的初步尝试，客观地讲，建成后的香积寺存在着诸多不足。如由于施工上难以保证大殿斜梁在屋脊节点处的刚性连接，为保持大殿的屋架稳定，后期施工时加了一根钢拉杆；部分梁柱结合处也有后加的角钢加强连接；又有施工水平、经验限制，部分隐藏节点构造没有做好，外表粗糙等。造成这样结果的有许多原因，最主要的是因为施工周期过于短促（从设计到施工主体完成不到一年），施工队伍（含专业施工队伍和传统木施工队）的管理和施工水平亟待提高，将现代木结构运用于传统的宗教建筑设计与施工之中，尚存诸多问题，需得到进一步解决，我工作室将在柳州开元寺、余杭超山风景区北苑建筑群项目的设计中逐步改进。我们坚信，随着国际交流的日益深入，采用先进的设计理念和施工技术，胶合木结构必将创造出新的现代建筑文化。

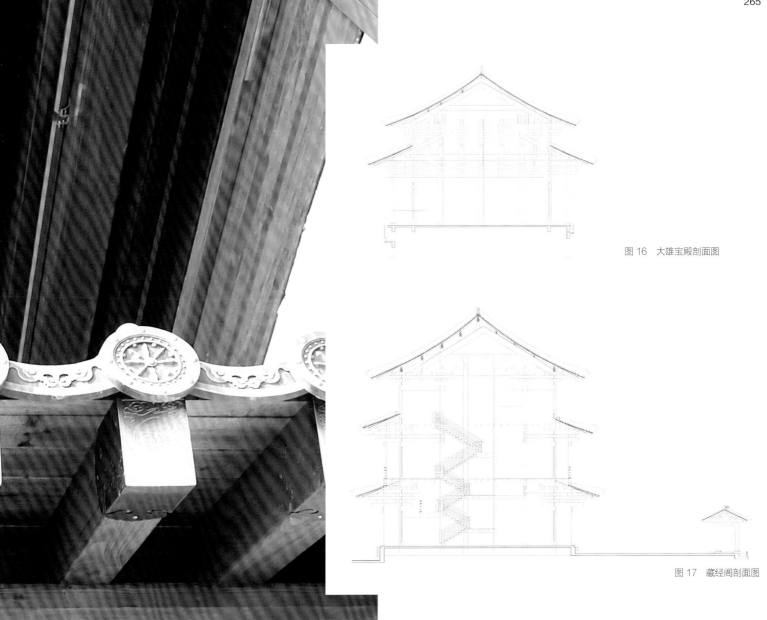

图 16 大雄宝殿剖面图

图 17 藏经阁剖面图

Grass frame
It is a roof beam, loading weight of the roof. In Chinese ancient architecture, such components are hidden above the ceiling. In Xiangji temple, four halls used the frame (figure 16, 17), which hidding in the oblique beam and smallpox, creating a traditional concave curved roof.

Reducing environmental damage to the gulam
Gulam will affect by external environment, especially the rain. Although it is embalmed, its section does not get this performance. As a result, the purlin section in the temple are claded with the copper, reducing the rain saturating the wood., presenting the intriguing detail design as well.(figure 18)

4.Conclusion

Glulam create a space and modeling detachment from the traditional religious buildings, making people feel the charm of religious buildings more deeply. The Xiangji temple is a preliminary attempt of glulam structure, objectively, there are many shortcomings in the project. Such as rigid connection of the hall oblique beam on the roof, we add a steel rod in the construction to maintain the main roof stability; there are steel rods in some beam-column junctions too; Due to the construction level, experience limitation, part of the hidden nodes structure was not perfect. There are many reasons for these difficulties, the main one is the short construction period (from design to construction subject completed in less than a year) and the level of management and construction team. There are a lot of problems of how to use modern timber structure of traditional religious buildings, The LIU Jie studio will be improved in the design gradually in the Kaiyuan temple (in Liuzhou) and ChaoShan scenic spot (in Yuhang,Hangzhou). We firmly believe that with the increasing development of international communication, advanced design idea and the construction technology, glued wood will create new modern architectural culture.

Photo Credit, the author

"土·木"再释
——深圳隐秀山居建筑设计中的"土·木"运用与实践[①]

Re-Considering "Earth" & "Wood"
Application and Practice of "Earth & Wood" in the Architectural Design of the Castle Hotel-Genzon in Shenzhen

作者：王兴田 | 校译：张亮 东鸿

在椰林树影的沙滩上，在白雪皑皑的山麓中，在茂盛葱郁的雨林里，在广袤无垠的草原上，休闲度假酒店总会以锦上添花的姿态出现在如诗如画的自然景致中，令人沉醉难忘。服务于休闲度假的酒店建筑与所处的山形水势、林木景致融为一体，并赋予人文内涵，唤起宾客情感上的共鸣，始终是休闲度假酒店设计的根本。隐秀山居地处景色怡人的岭南地域，青山碧水间，酒店若隐若现，推窗即景，大自然触手可及。建筑师以取之天然的"土·木"为设计材料，以营造建筑与自然和谐相融的环境为目标，提炼融合地域属性和文脉特征，形成"生于斯，长于斯"的建筑特色，使宾客与周围环境达到物我相契、情景交融之境。

In beautiful landscapes, a resort hotel designed well can complement the wonders of its surroundings. Such an effect can be realized in the shadows of a palm-tree covered beach or amidst the contours of a snow-filled mountain slope. It can be achieved amidst the canopy of a lush rainforest, or between the gentle slopes of an expansive grassland. Regardless of where, the core of resort design has always consisted of architecture integrated with surrounding mountains, water and trees, alongside content that produces an emotional response.

"土·木"再释
Re-Considering "Earth" & "Wood"

　　隐秀山居以"在地"为原则,依照原本的山形水势进行设施布置,酒店建筑一边面向球场、庭园、自然景观,另一边面向城市,通过种植造景使热闹逐渐向宁静过渡。基地内泳池、山水及文化"老宅"的植入,构成恬静的庭园。各功能区以大堂为中心联系成一体,西侧为商务会议场所、东侧为宴会厅及各式餐饮空间;为更有效地利用好基地,设计将康体娱乐区、后勤服务区以及设备空间置于地下,通过天井导入自然光和通风,同时利用台地开敞面布置主要功能空间,与室外绿化相连,赢得良好的景观视野。

The Castle Hotel is located in the picturesque Guangxi-Guandong region of "Lingnan," its form shimmering amidst green mountains and clear water. Here landscapes beckon beyond every opened window, and nature is always within grasp. The architects used natural earth and wood for building materials and set off to pursue a goal of harmonious consistency between architecture and nature. With this orientation, they refined regional attributes and cultural or historical contexts to produce architecture locally "born and raised." The result allows guests to easily-explore the surrounding environment.

"土·木"再释
Re-Considering "Earth" & "Wood"

纵横布局是东方建筑空间组织中常用的手段,古代周王城九经九轨的城市格局,井田制的乡镇概念等都是纵横格局的经典运用。"纵横"手法看似简单重复,实则韵味无穷。纵横交错间妙趣横生,带来无限可能,使建筑不仅能在大空间上延续伸展,每一个小的角落也可向自然延伸。隐秀山居运用纵横轴向布局策略,使酒店空间在纵与横、内与外中展开。建筑师匠心独运,在两侧延伸的公共空间中设置等级差别的多片段组合,摆脱了酒店狭窄地形上的空间局限,实现了处处皆风景的诗意生活。

隐秀山居以中央大堂空间为中心沿高尔夫球场面向东西两侧有节奏的转折展开。"纵、横"的空间变换,交错引向深处,向自然面伸展。受地界及高压线的限制(界限狭窄的基地东面、南面有高压线穿过,西侧、北侧是生态用地),一字形酒店布局成为唯一可能,建筑通过两个转折将纵向空间串联在横轴上,在横轴空间上移动和延伸过程中,纵向轴空间有节奏地出现,把自然直接引入廊道中,借鉴传统园林的廊道意象表达,自然在忽隐忽现中变得有趣、悠然,也使人从心理上缩短了300米冗长的横向动线轴。

The Hotel Genzon uses the principle of "grounding in locality" to integrate its buildings in adherence to the topographical forms of adjacent mountains and according to the course of surrounding water features. Hence the hotel is constructed with one side facing a natural landscape, athletic field and garden. Another side gazes out towards the city. As a result the plan creates a transition for visitors from a place of urban excitement towards one of natural tranquility.

At its base swimming pools, landscape features, and a traditionally-designed house constitute an idyllic garden. Various functional areas are connected together via the central lobby. At its west is a business conference center. Banquet rooms and a variety of dining facilities are situated at its east side. In order to better make use of the base, the design has situated the entertainment and logistics service areas as well as the equipment space underground. Natural light and ventilation is provided to these areas through the open space overhead. This area then serves as the site for essential hotel functions, and is connected to the outdoor landscape, producing a scenic view.

"青山看不厌，流水趣何长"，中国人的自然观是让身心融于青山绿水中，寻得超然自在。建筑也不例外，它们源于自然，通过当代技术的提升，又恰如其分地回归到自然中，自然而然、浑然天成。"土""木"是大自然赐予我们的恩惠，建筑师把"土""木"作为建筑特色融入到山水之中，通过从整体到细节的设计，可以感受到对自然资源的尊重和合理利用。

A vertical and horizontal layout is commonly used in the spatial organization of Eastern-architectural space. The ancient urban grid-pattern of "Nine Horizontal and Vertical Paths" from the Zhou Dynasty, and the township concept of the "Well-field System"[2] represent classical applications of this form. These forms seem simple but in fact contain infinite potential variations. The crisscross pattern creates a wild array of charming possibilities, and allows buildings to extend across a large expanse as well as permitting the natural expansion of every little corner of construction. The layout strategy adopted by the Castle-Hotel makes it possible to extend its space vertically and horizontally, internally and externally. In addition, through an original intervention by the architects, the hotel evades spatial constraints. This is achieved through the installation of unique segments at different grades in the public space extending from its two sides. The result is a scenic and poetic experience.

[2] This system established a common field for taxation in the center of eight adjacent plots, and is symbolized by the character for "well" in Chinese (井).

"土·木"再释
Re-Considering "Earth" & "Wood"

The Castle-Hotel spreads rhythmically to the east and the west of its central lobby along a golf course. Through vertical and horizontal transformations, the structure successively expands into the natural surroundings. The structure is restricted by high-voltage power-lines which pass through the south and east sides of the site. Ecological land sits to the north and the west of the site. On this basis, a straight-line layout is the only option. The vertical space is connected to the horizontal axis through two bends. It appears to move and spread rhythmically alongside the axis. Nature is introduced into the horizontal corridor through images of traditional gardens. Hence nature flickers throughout the 300 meter-long horizontal axis. This has the effect of shortening its perceived distance and results in a leisurely and amusing passage.

当代木构建筑以胶合木为建筑材料,重量轻、强度高、美观、加工性能好,木结构更是具有绿色环保、节约能耗、抗震性能好、施工快捷等特点,其与人类的亲密性和亲近感是其他任何建筑材料难以替代的。几千年来中国建筑体系是以木构为框架的结构体系,尽管随着建筑技术的不断进步,新型建筑材料层出不穷,但与人最为亲近的天然材料情结却始终未变。随着生态低碳生活的观念倡导,木构材料作为"绿色"新宠,重新回归久违的建筑舞台。当代胶合木与其他建筑材料相比,有着无法比拟的低碳优势。木构建筑从采伐、运输、加工、建造到使用、解体、或被再利用的全部生命周期中,能源的低消耗和碳的低排放令其环保性能尤为突出。通过多元材料和复合技术的综合处理,木材可以防火、防腐、防潮,木的使用寿命和承重性等种种问题都能迎刃而解。木构在建筑之中的使用充分体现了人、建筑、自然的和谐统一。

Chinese people commonly view green mountains and clear water as places to find comfort and transcendence for mind and body. As the Chinese saying goes, "green mountains do not annoy, and flowing water is always a joy." Buildings are no exception. Their source is in nature. And through the advance of contemporary technology they can be integrated appropriately back into nature. "Earth" and "wood" are bestowed by nature. Through overall and detailed design, architects can insert "earth" and "wood" within a landscape as architectural elements. In doing so, they can respect and rationally use national resources.

随着技术的不断进步和发展,建筑不断向新的高度发出挑战,奢华与浮躁之风影响着人们对建筑价值观的判断。或许一切事物都会沿着一定的规律自然循环,当人们在钢筋水泥中得到了高度和规模的心理满足,物质的炫耀后,随之而来的是对回归自然的渴望,于是,源于自然的土木有了更高层次的回归。当代木构建筑在北美、北欧和日本等地已广泛运用。在我国,由于近现代林木过渡无序的采伐,破坏了森林资源的可持续发展,木构材料的运用和技术的研发在建筑史上一度出现了空白。

Contemporary wood-structure construction can use glued laminated timber, also known as glulam, for construction. This material is lightweight, high strength, aesthetically pleasing and easily adapted and improved. In addition, it is environmentally friendly, energy efficient, conveniently assembled, and has been shown to have excellent seismic-performance. As a result, it is highly compatible with human uses, to a degree difficult to match with other materials. For thousands of years, Chinese architectural systems have been based on wood-frame structures. This has been the case in spite of the unceasing improvement of building technology and the emergence of new building materials. All along, people's preference for natural materials has remained. With the rise of low-carbon ecological living models, wood has been recognized as "green" and has regained its long-lost recognition as a premier building material. Compared to other materials, contemporary glulam has the unparalleled advantage of being low-energy-consuming and having low-carbon emission. As a result, its environmental qualities are superb. Because of its integration of multiple materials and through its use of composite technology, the material can resist fire damage, corrosion, and moisture-damage. In addition, it greatly exceeds standards for longevity and limits on load bearing, resolving all concerns. The use of this wood material in construction hence is ideal for forging a unity between people, architecture, and nature.

笔者在日本留学、工作期间对木构建筑有一定的认识和了解，从"源于自然、顺应自然、回归自然"的设计理念出发，在项目策划设计之初，便想营造一处回归自然、"择木而栖"的优雅山居。只是在当时国内的现行规范下，三层以上的建筑缺少木构实现的可能，在与行政部门多次沟通探讨后，只能保留木构主入口雨篷的尝试，也算第一次实现了融于自然的木构情结。在整个项目设计过程中，建筑师和结构工程师的密切配合、国外先进材料和技术的引进以及各方的共同努力，成就了当时国内跨度最大的木构雨篷。

"土·木"再释
Re-Considering "Earth" & "Wood"

With the unceasing improvement and development of technology, architecture always is reaching new levels. At the same time, a standard of what is vogue and changing fads influence people's judgment of architecture. Trends have a naturally cyclical form. Following a period in which psychological satisfaction is derived from reinforced concrete and ostentatious displays of excessive heights and large-scale buildings, people will return to yearn for nature. Thus naturally sourced earth and wood will return to occupy a higher stage. Contemporary wood building and construction has been widely implemented in places such as North America, North Europe, and Japan. In China, as the sustainable development of forest resources has been undermined due to excessive logging in the modern period, the use of wood materials and research and development into wood technology have been neglected.

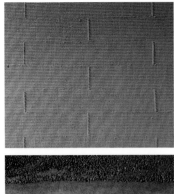
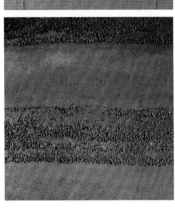

隐秀山居入口雨篷采用花旗松胶合木的重型木结构，其给人的震撼来源于木材本身就是一种最质朴最真实的表达；同时，雨篷采用"柳暗花明又一村的"布局方式，于密林深幽的桥木林后，穿过一条曲径通幽的林荫小道，眼前豁然开朗，一个造型独特的木构雨篷便映入眼帘，其下"内聚"的流水、叠石、灌木相互交渗，渐渐在环境中隐去。雨篷设计通过对结构完整的逻辑诉说表达传统韵味，同时也兼顾当代感和技术展现。木构雨篷位于建筑南端，平面由中轴对称的不规则六边形构成，纵横形成 27m 大跨度空间；立面沿建筑中轴对称，屋脊位于对称轴上，呈倾斜向上的动势，顶部最高点距地 15 米，最低点距地 8 米。雨篷结构主体由 5 榀非规则木桁架斜向支撑，在屋脊处对接，各榀桁架之间的上下弦由木梁拉接，形成稳定的屋面桁架体系。木桁架之上，成组排列着木构檩条，巧妙承担着雨篷屋面的覆盖和出挑。雨篷整体构造逻辑清晰，节点构造精巧细致，尽显木构"天然去雕饰"的质朴优雅。

"土·木"再释
Re-Considering "Earth" & "Wood"

During overseas study in Japan and in professional practice, the author has gained a level of knowledge and understanding of wood architecture. Starting from a design concept of "derivation from nature, compliance with nature and return to nature", the architect initially intended to create a "return-to-nature" themed bucolic and elegant mountain residence during the beginning stages of planning and design for this project. However, according to existing domestic requirements, it was impossible to construct a wood structure in excess of three stories. After several interactions with the department of administrative affairs, the timber awning at the main entrance of the structure was the only wooden component retained from the original plan. However, this feature realized the designer's ambition to build a wooden structure in harmony with the natural landscape. The overall project involved close cooperation between the architects and structural engineers. It also required the introduction of advanced materials from overseas as well as new technology. Through combined efforts of the various concerned parties, the largest-spanned awning in China ever was successfully assembled and built.

建构是建筑师的视觉呈现之道,旨在强化那种属于艺术领域的现实体验,也就是对建筑中的形式与受力关系的体验。因此,作为一个不可触摸的抽象概念,结构通过建造得以实现,并且通过建构获得视觉表达。[①]隐秀山居的建构是结构、构造、建造三位一体的集合,是一个全过程的综合反映,其本质是将空间、结构、形式、功能尽可能完整地融合为一体。因此,在木构建造过程中,讲求结构的合理性、清晰的构造逻辑以及形式的朴实质感,并将三者完全呈现出来,形成统一的建构,让建筑的表现更富原始魅力。

①[美]爱德华·F·塞克勒:《结构、建造、建构》,1965年。

The awning at the entrance of the Castle-Hotel uses heavy-duty Douglas fir glulam. It is the wood itself that shocks people. It is a most rustic and honest form; at the same time, the awning brings to mind a theme of "hope in the midst of darkness and confusion." The unique wood awning comes into focus suddenly after walking through a winding tree-lined trail abutting a dense forest. Underneath it, the combined effect of water, folded rocks and shrub features interact with each other, and gradually disappear into the unfolding environment. The awning's complete structure expresses a traditional charm while also referencing contemporary themes and technology. The wood awning, located at the building's south, is composed by an axis symmetric irregular hexagon. This forms a large-span 27 meters wide and long. The awning starts from the building's central symmetrical axis, from which points its roof ridge rises upward. The top of the roof is 15 meters above the ground and at its lowest point is 8 meters high. The awning structure is supported by 5 pieces of irregular wood truss which are jointed at the ridge. The top chords and the bottom chords between the trusses are pulled together with the wood beams to form a stable roof truss-frame system. The wood purlins are neatly arranged in a clever manner on top of the wood truss to bear the load of the covers and the cornices of the awning's roof. The overall awning structure features clear logic and delicate node details. It excellently represents elegant rustic timberwork without excessive decoration.

　　肌理涂料是感知的主干,其他覆层材料都是参与设计公共空间界面的片段组织。休闲度假酒店重在丰富界面,缩小尺度,消除单调,营造休闲欢快体验的感受,而南"动"(公共)北"静"(私密)的思考,落实了材料的物性和传达的"表情",也许是一种混合的解理性,或给人一种陌生的新鲜感,或是建筑"在地"与生俱来的"地域性"。

Construction is the way in which architects render a vision. It aims to strengthen the sort of experience of reality belonging to the field of art, that is, the experience of the form and the internal forces within a building. Therefore, as an intangible abstract concept, the structure is realized through its construction and obtains a visual expression through being built.1 The construction of Castle-Hotel features the integration of structure, construction and building, which is a comprehensive reflection of the entire process. Its very nature requires the integration of space, structure, form and function as much as possible; therefore, the process of wood construction emphasizes structural rationality, clarity in tectonic logic, and honesty in form. These three components form a unified construction and enable the architecture to represent a more original charm.

隐秀山居酒店整体力求简约、舒展、飘逸，与"在地"环境相辅相成。建筑外墙底部采用珍珠绿石材，除开窗外，80%均采用近似大地土壤色的经济涂料，从色彩上弱化工业痕迹，从而与环境相融，通过调整涂料的黏稠度和厚度，工匠用金属手工梳整出横向的涂料肌理，既增添了手工艺的质感，让细节更富有情趣，又实现了低投入、低能耗，与当代低碳的生态观念不谋而合。酒店近似天然的外立面效果，营造出从大地泥土中生长出来的建筑之感，暗合中国循环、相生、互换与长久的自然观，让"土、木"回归，为建筑整体勾画出乡土古朴的色彩和质感，表达了建筑师回归自然之匠心。

远离都市的喧嚣，拥抱绿色的自然，倾听生命的声音，感受和谐的韵律，这就是设计者力求营造的给人的第一感觉。隐秀山居遵循绿色的建筑理念，以生态设计手法，运用天然木材与现代技术的结合减少建筑对环境的负荷，与山水对话，与自然交融，让基地中的一山一水、一草一树都充满了生命的气息，循着山水走到内心深处，去感受大自然的无限生机，去领略天地间的无尽禅意。

External appearance is a central form of perception, and other outer materials are all elements of the design of public space. The resort hotel emphasizes the enrichment of its interface, the reduction of scale, the elimination of monotony, and the creation of enjoyable, fun experiences. In the meantime, consideration of the "active' and public south part and the "quiet" and private north part informs the nature of materials and the expression conveyed. This can be understood as a sort of mixed rationality, perceived as a strange novelty, or seen as an expression of architectural locality born "on the ground."

The Castle-Hotel is simple, elegant, and is "from the earth" while at the same time complementary to the earth. The base of its exterior walls make use of green pearl stones and 80% of walls use economic paint with earth tones, which lessens industrial-chemical elements in the paint and better integrates itself with the surrounding environment. In addition, by adjusting the viscosity and thickness of the paint with metal combs, painters not only improve the aesthetic quality and make details richer and more interesting. They also save resources and reduce energy consumption, hence adhering to contemporary low-carbon ecological standards. The natural-like appearance of the hotel façade creates the sense that architecture is almost growing from the earth, and is in accordance with traditional Chinese views of natural circulation, coexistence, mutual exchange, and longevity. This allows the return of "earth" and "wood," and lays out a tone and texture of ancient simplicity for the architecture. It represents the architects' innovative turn back to nature.

The architects sought to instill the following first-impressions in visitors to the site—an awareness of escape from the noise of the city, a sense of embrace of the natural environment, the hearing of sounds of life, and the feeling of a harmonious pulse of energy. By adhering to principles of green architecture, Castle-Hotel has adopted ecological design methods, and used the combination of natural wooden materials and modern technology to reduce the burden of buildings on the environment. This allows the site to interact with the landscape, blend in with nature, and enables the mountains, water, grass and trees on the site to be animated with the breath of life. We can walk through the landscape to our hearts' content, feel the boundless vitality of nature, and taste the infinite Zen between heaven and earth.

成都毗河项目基本信息
项目地址：成都市新都区毗河南岸（曲景路段）
项目功能：原为集文化沙龙、展示、休闲一体化的多功能中心，
　　　　　现打造成特色餐厅（里面整合了部分文化交流的内容）
建 筑 师：成都草晖设计顾问有限公司　康宁
结构工程师：苏州拓普建筑设计有限公司　杨春梅
建筑面积：2463m²
建成时间：2011年12月

成都毗河项目经验分享
Chengdu Pi River Project Experience Sharing

作者：周金将 郑小东 徐升阳 王永兵 | 校译：高瑜 东鸿

简介
Introduction

本工程位于四川省新都区毗河南岸，属于政府重点打造的毗河沿岸文化走廊的一部分。该项目为大跨度的胶合木单梁结构，胶合木单梁跨最大 16.7m，整个建筑面积 3560.2m²，胶合木用量约 235m³。

This project is located on the south bank of Pi River in Xindu district, Sichuan province, which is part of cultural corridor along the bank, invested by government. This case is single beam glulam structure with the largest span up to 16.7 m, whole area covered 3560.2m2 and with 235m3 glulam.

设计

该项目丙类建筑。结构设计基准期为50年,重要性系数为1.0,抗震设防烈度为七度,设计基本地震加速度值为0.10g,水平地震影响系数最大值为0.08。建筑场地类别为Ⅳ类,设计地震分组第一组,场地特征周期为0.65s,结构耐火等级为二级。

重木结构梁采用胶合木(GLULAM),等级为Spruce-Pine 20f-EX。柱采用Douglas, fir-Larch(D.Fir-L),等级Ⅰ级,木结构含水率低于15%。所有直接暴露于室外木构件及无防水层的露台木结构构件及与混凝土直接接触的构件在施工现场采用ACQ进行防腐处理。金属连接件采用钢材制作,材料为Q235B,钢材的抗拉强度实测值与屈服强度实测值的比值不小于1.2,钢材有明显的屈服台阶,伸长率大于20%。锚栓材料为Q235B,螺栓采用4.8级普通螺栓。

结构整体计算采用SAP2000V9.16有限元计算软件,模型如图1所示(选取体量最大的餐饮大厅为例)。梁的截面形式有250mm×600mm和250mm×800mm两种形式,跨度最大的达到18.53m。柱截面亦有两种:内部木柱的截面为250mm×800mm,外部木柱为变截面,柱顶截面为250mm×840mm,柱底截面为250mm×495mm。

Design

It is Class C building with 50 years design service life. Building importance category is II, seismic precautionary intensity is 7 degrees, basic design earthquake acceleration value is 0.10g, maximum value of influencing coefficient of horizontal seismic action is 0.08. site earth classification is Class Ⅳ wit earthquake group of Group I and characteristic period is 0.65s. Fire rating is class II.

Primary beam is glulam with grade Spruce-Pine 20f-EX. Column is Douglas, fir-Larch(D. Fir-L), grade I. moisture content is below 15%. All wood elements which are exposed or contact with concrete are with ACQ treatment. Steel connector is Q235B, the ratio of tensile strength and tested yielding strength should not less than 1.2. the steel shall have obvious yielding plateau with plastic elongation greater than 20%. Bolt is class 4.8 with material Q235B.

The structural global calculation is done by finite element software SAP2000 V9.16, refer to image 1(selected model is canteen with largest span). Beam section is 250mm×600mm and 250mm×800mm, the largest span is 18.53m. column section with two types: internal column's section is 250mm×800mm and external column's section is tapered with 250mm×840mm on top and 250mm×495mm on bottom.

主要节点形式在软件中的假定如下：由于梁柱节点处梁柱截面在框架平面内的高度均较大（分别为600mm和840mm），并且在两侧有木构件支撑，又通过内嵌钢板螺栓连接节点可以实现弯矩的有效传递，因此在软件中假定为刚接。柱脚节点（柱脚部分截面为250×495）假定为刚接；梁梁连接节点假定为铰接。

根据SAP2000内力计算结果对各个构件进行强度、稳定性、变形验算，结果表明模型计算所预先假定的构件截面均满足要求。在此基础上，设计各个梁柱、柱脚及梁梁连接节点。

Connection hypothesis as: connection between column and beam is semi-rigid, connection between column and foundation is rigid and connection between beams is pin. Strength, deflection, and stability are calculated by SAP2000, results are within design criteria. And all connections are calculated base on global calculation.

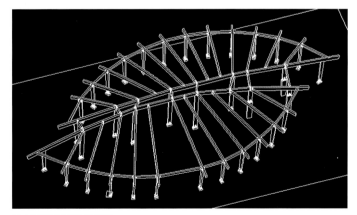

图1 SAP2000 计算模型
Illustration1 SAP2000 calculation model

加工
Processing

施工图完成后开始工业详图设计，为使拆分精度受控，以及能及时发现后续安装过程可能存在的问题，建立了3D模型，在电脑中检查各个木构件与钢连接件之间的连接效果。部分柱脚、梁柱、梁梁的节点连接如图2、图3、图4、图5所示。

Detailed processing drawings are started after construction drawing. With the purpose of disassembling accuracy control and timely detection of problems during the following installation procedure, 3D models are constructed on computer, where connections of timber components and steel connectors will be checked. Some pillar bases, beams and beam-beam connectors are illustrated in Illustration 2, 3, 4, and 5.

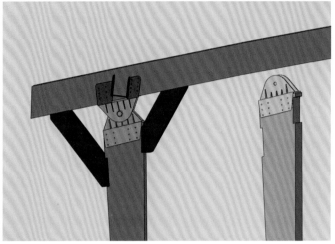

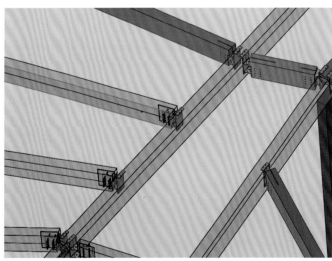

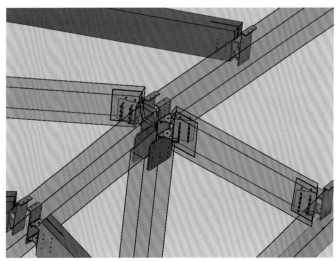

图2　柱脚连接节点
图3　柱梁连接节点
图4　梁梁连接节点
图5　梁梁柱连接节点

Illustration 2　pillar base joint node
Illustration 3　joint node of pillar and beam
Illustration 4　joint node of beams
Illustration 5　joint node of beam-beam and pillar

经过工业拆分设计之后,图纸送到加工中心加工构件。该项目柱子最大尺寸为300×800×9375(外围柱子为变截面,下端截面为300×495,上端为300×800),主梁最大尺寸300×800×14150,弯梁最大尺寸300×600×18530。在加工中遇到的问题就是现有的生产线不能完全满足该项目胶合木加工的要求,后来加工中心通过讨论研究,把一条立式的胶合木生产线改成了卧式的,这样就实现了弯梁胶合木的加工。下图6为加工中心立式的胶合木生产线。

Drawings of industrial process and component design will be delivered to machining center for component processing. The largest size of pillar in this case is 300×800×9375 (the peripheral pillar is variable cross-section, the bottom section is 300×495, the upper one is 300×800), the largest size of main beam is 300×800×14150, and the largest size of curved beam is 300×600×18530. The problem during the processing is that the current production line would not meet processing requirement of glulam in this case. According to relevant discussion and research, a vertical glulam processing line is transformed to horizontally line, which will implement processing of curved glulam. Illustration 6 below is vertical glulam production line in machining center.

图6 加工中心立式胶合木生产线
Illustration 6 vertical glulam production line in machining center

施工

该项目现场的安装遇到的挑战也是多方面的。首先原有砼部分已施工完成，后来经过复测，部分柱子基座的位置有偏离，需重新调整位置。其次现场入口处有限高，起重吊车无法正常进入。结合现场的实际情况，在胶合木进场前我司就进行了针对性的调整。对于偏位的柱子基座，重新放线后确定位置制作模板，专业植筋后重新浇筑砼（见图7、图8所示）。针对现场入口处限高的问题，在胶合木进场后，首先用大吊车将中间位置的胶合木柱吊装到位，再将小吊车直接吊入场地内进行轮转吊装作业（见图9、图10所示）。

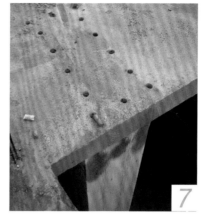

图7　二层板上专业打孔
图8　二层板上专业植筋
图9　大吊车吊装大梁
图10　利用大吨位吊车将小吊车吊入场地内
Illustration 7　professional drilling on floor 2
Illustration 8　professional planting bars on floor 2
illustration 9　heavy crane is hoisting the main beam
Illustration 10　make use of the heavy crane to hoist the small crane into site

Construction

There are challenges on field installation in many aspects as well. Firstly, the construction of concrete part is done, where some pillar bases are out of position and require adjustment. Secondly, there is limited height on entrance, where crane-truck would not enter normally. According to the field situation, we finished specific adjustment before glulam's ready on site. The pillar bases out of position are reset out. We manufacture template and replace concrete after professional planting bars are done (please see in illustrations 7and 8). To deal with the limited height at entrance, heavy crane is operated to hoist center-position glulam pillar in place and then hoist the small crane into site to take turns to work with the heavy one (please see in illustrations 9 and 10).

图 11 现场结构完成后内部照片
Illustration 11 interior view of completed structure construction

图 12 现场结构完成后外部照片
Illustration 12 appearance of completed structure construction

　　胶合木安装过程中经历了雨季，现场对于胶合木的保护异常困难，在胶合木端头用塑料薄膜封住。在工程项目组共同努力下，结构安装顺利完成，并经过了甲方、监理、质检单位的顺利验收。现场结构安装完成后照片如图11、图12所示。

The glulam protection is really difficult, for the installation procedure is just in the rainy season. The solution is to seal up both ends of glulam with plastic membrane. Construction installation is completed successfully under the joint efforts of the project team, which wins acceptance of client, site supervisor and quality supervisor. The completion of structure installation is shown in illustration 11 and 12.

图 13　装修完成后外部效果照片
Illustration 13　appearance effect of completed decoration

该项目完成后作为当地餐厅及文化展览相结合的建筑来使用，利用木结构的自然环保，与周边的河流、景观达到完美的统一，使每个来就餐的人都能心神愉悦。在其间不时举办各种文化沙龙、名画展览等，更使其功能特色得到进一步的发挥。现场部分效果如图13、图14、图15所示。

图 14 装修完成后内部效果照片一
Illustration 14　interior effect of completed decoration

This timber architecture is the nature-friendly complex of local restaurants and cultural exhibition, which reaches the great harmony with surrounding river and landscape and makes visitors comfortable. Its functional features will be further used in the various cultural salons and exhibitions of great pictures frequently. Some effects of site are shown in illustration 13, 14 and 15.

图 15 装修完成后内部效果照片二
Illustration 15　interior effect of completed decoration

小结 / Summary

最终简单分析一下该项目的经济影响。成都这个地方在国家西部大开发战略中有着举足轻重的作用，借着西部大开发的机遇成都这些年的经济发展速度在全国来说也算是比较领先了。加上自古以来就有"天府之国"的美誉，对于成都的经济、文化等各方面的快速发展，多数企业都跃跃欲试。新都区位于成都的北边，近些年成都市又大力推行北改工程，就是要把新都、青白江这些区域与主城区连成一体发展。该项目木结构部分建筑面积为2463m^2，施工内容包括主体结构、屋面装修，造价为492.6万元，约合每平方2000元。

借着国家支持推广低碳节能环保建筑的东风，现代木结构特别是胶合木结构这种结构体系在未来的中国建筑市场将占据越来越大的份额，作为行业中的一员，我们对木结构的未来充满信心。

In the end, we will give a simple analysis about the economic influence factors. Chengdu plays a pivotal role in China's western development strategy policy, which helps Chengdu's economy develop faster than most of the cities in China. Additionally, Chengdu has a reputation as Land of Abundance with rapid development on economy and culture, which make most enterprises eager to take challenges than just standing-by. Xindu district is located in the north of Chengdu. These years, Chengdu government vigorously promotes Reform Northwards Project, with the purpose to unite remote areas, like Xindu district and Qingbaijiang district, with central downtown for unified development. The construction area of timber structure is 2463m2in this project, with construction contents of main structure, roofing decoration, which costs RMB 4.926 million Yuan and about RMB 2000 Yuan per square meter.

With the help of national policy on promotion of low-carbon, energy-saving, and environment-friendly buildings, modern timber structure, especially glulam structure system, will play more and more important role in domestic construction market in China. As an enterprise in this industry, we have full confidence in timber structure's future.

云建筑
——万科青岛小镇游客中心诞生记

Cloud Building
Vanke TsingTao Pearl Hill Visitor Center

作者：詹晖　汤玉辉　庄寅麟　｜　校译：詹晖　东鸿

　　在钢筋混凝土丛林之中，木材作为一种结构材料在当今已是个稀罕物。既然是个稀罕物就应有其特别可爱之处，否则稀罕物就要慢慢退出世人的眼界了。可幸之处是万科作为国内数一数二的房地产开发商（估计在全球也是排得上座次的，无考证），继吉林万科城售楼处项目采用木建筑之后，再次选用木结构并决定由上海思卡福来担纲万科青岛小镇迎客中心项目的建设。眼下项目已完工两年有余，这两年多来，该项目已拿到数十个大大小小的奖项，以兹证明它还是的确有些可爱之处。下文就和诸位读友浅尝一下这位颇有些可爱的稀罕之物。

In concrete and steel jungle, wood as a kind of building material is a rarity in recent decades. Now as a rarity it should have special lovely place, otherwise the rarity will slowly withdraw from the secular vision.

Luckily is that Vanke as one of the biggest property developers in China, after the Jilin Vanke city sales center adopting wood framing, Vanke selected wood structure again and selected SKF to undertake "Vanke TsingTao Pearl Hill Visitor Center" project. The project has now been completed more than two year, during these time, it has got dozens of local and global awards which indicate it has some lovely places. The following are some introduction about this building.

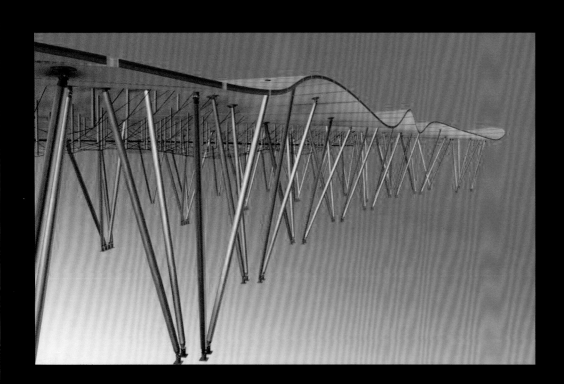

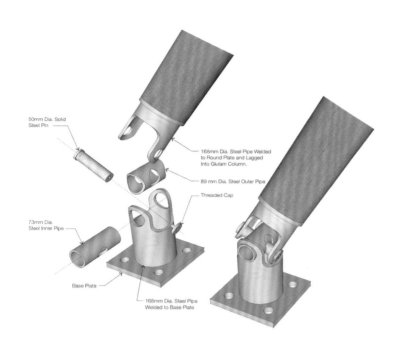

为了便于大家更好的了解这个云建筑（姑且这么叫，至于为何如此称呼，且看下文），这里先占用大家的一点时间介绍一下它的祖籍背景。云建筑诞生于青岛西海岸滨海大道小珠山南麓，其西北与青岛珠山国家森林公园相邻，整个开发项目三面环山，南向看海，对于建设青岛小镇的构想主要来自于结合原地固有的天然景观，打造一个全生态绿色环保的可持续发展社区。出于对这块土地的珍惜，开发团队经过数次跋山涉水进行实地勘测，并以追求居住的原生态精神，遵循保留原貌、依山就势的原则进行项目开发。项目总建筑用地一千多亩，绿化用地四千多亩，总建筑面积约一百万平米。而云建筑作为该项目的点睛之笔（前期作为售楼中心，后期作为游客中心），坐落于地块前端进入项目腹地的必经之路口，依山而立。

For a better understanding about this cloud building (let's just called so now, as to why such call, please see below), let's take a little time to introduce its background. Cloud building is located in the west coast of TsingTao, pearl hill foothill, the northwest adjacent to the national forest park. The idea of develop the whole project (TsingTao town) mainly comes from the combination of the inherent natural landscape to create a fully sustainable development and ecological green community. For this purpose, the development team traveled to field survey, followed the principle of keep original for the project development. The cloud building as a highlight of the whole project is located in the entrance to the town and stand up to the mountain.

当然好山好水的先天条件并不代表就一定能产生好的项目,孕育过程亦是相当之重要。为此项目特别聘请了顶级法师－大洋彼岸的BCJ（Bohlin Cywinski Jackson 建筑事务所）担纲建筑设计。洋法师亲临现场巡视一番,当夜便在塌下挥毫,寥寥数笔,便勾勒出了云建筑今生的雏形：林回路转,曲径通幽,一片浮云,依山而立,似闲庭信步,或曼妙秀舞。次日会审众人见此案稿无不拍手称赞。而后建筑内部的设计通过有序通透的开放空间,将功能区域进行系统分割,从而形成销售、展览、休憩、购物等功能空间的布局。

至此,云建筑它今生的模样便在洋法师的笔下成形了。然而模样之于外,于内之骨骼脉络同样不容小觑。也许是今世注定它的血统中不能少了洋味,枫叶之邦的老法师 Gerald Epp 被请至担纲结构设计。不愧是木结构界的翘楚,什么奇形异状在他的手下都能整的巧夺天工：万向连接件、树状锥形胶合柱、规格材拼接曲面板、钢拉索、短立柱,最后由 93 根锥形胶合木柱支撑起整个面积达 2600 平方米的三维曲面浮云造型的木屋顶,堪称一绝！

Such a place with good scenery doesn't link to a good building directly. For a good design, we invited BCJ (Bohlin Cywinski Jackson architects) to do the architecture design. After visiting the job site, the architect draw the outline of the cloud building: forest turn with winding street, a cloud is situated at the foot of a hill, it may wish to take a stroll, or graceful dance show. Everyone applauded when he show us with the concept design on the meeting the next day. And per the plan and interior design later, the building room was divided into some open spaces with different function areas such as sales, exhibition, recreation, shopping space.

After the completion of architecture design, we invited Gerald Epp from Canada to act the structural design. The architecture is very fantastic but it is a big challenge for wood structure design: 3D curved roof, not enough shear wall, column with universal direction…and etc, but all difficulties were overcome by the structural engineer: the universal connectors, tree cone column agglutination, curved plate stitched by SPF, steel cables, short web columns, and finally the whole curved roof which is almost 2600 square meters is supported by the 93 pieces wood columns together with steel cables and web columns, absolutely splendid!

　　好了，模样和骨骼都已孕育完成，接下来便是接生落地的活了，这也是我们的拿手好戏，SKF 上海思卡福的师傅们通过近二十年的木结构实战经历积累下了也算是丰富的经验，建厂房、加工曲面板、立柱子、支墙体、吊装屋面、装短柱，拉缆索……虽说是经验丰富，但大多也是本土做法，遇上这洋娃娃还是经历了不少坎坷，研究、讨论、请教、创新、试验，历经了多少个不眠之夜，终于盼到它诞生了。

At this point, all design had been done, the next was fabrication and installation which are SKF specialties, Shanghai SKF construction Co.Ltd has accumulated a wealth of wood structure construction experience through nearly 20 years practice. for this project, we built workshop near the job site, fabricated the curved plates, installed vertical columns and walls, lift the panels, pull the cables with web columns…… Though we are experienced with wood structure, but this building was still a challenge to us, after many sleepless days and nights with research, discussion, innovation and test, eventually the cloud building was born.

落地以来，云建筑得到了客户、开发商、设计师、建造商及其他社会各界的一致认可，里里外外获奖数十个，这倒也没辜负当时开发、设计、施工等等诸位的良苦用心，这也便是我们一开始说它有些可爱的依据。然而回味之余，我们想从它的被认可推测出几条关于木建筑的思考，以供探讨：

Since the completion, the cloud building has won the applause of the clients, developers, designers, builders and others with dozens of awards. with the success of this project, we want to share some options about wooden structures in China for discussion:

作为拥有几千年建造、使用历史的木建筑，现如今隐于主流建筑市场有其特定的社会背景。作为当下木结构建筑的从业者，一定要认清它与主流的钢筋混凝土建筑的不同之处，分析需求，精准定位，发挥特色，方能赢得认可，站稳脚跟，进而扩大市场。

要做出好的木结构建筑，设计是灵魂。现阶段因各方面原因国内木建筑的设计力量还不够强大，优秀的设计大多还要依赖国外设计师。但只要我们专注，广学博览，多联系实际，发挥特色，就能涌现出越来越多的优秀木建筑设计师和优秀的木建筑作品。

打造优秀的木结构建筑，施工是关键。一个建筑作品最后呈现在世人眼前时被认可与否，施工质量把控起着重要作用，这会直接影响到建筑的外观、使用性及安全性。相比主流的建筑形式，木建筑的成败与施工质量的好坏相关性更高，精雕细琢，精益求精是做好木结构建筑的基本保证。

1) With a thousands of years build and use history in China, now the wood structure's situation -out of the mainstream of construction market has its particular social and economic background. As the practitioners of wood structure, we must recognize it from the mainstream of concrete or steel structure, focus on the demand analysis, accurate positioning, strengthening characteristics, and then win the recognition and expand the market.

2) To build a good wood structure, design is the soul. At this point because of various aspects of reason, the domestic wood building design strength is not strong enough, good design mostly relied on foreign designers. But if we pay more attention, more learn, more exhibition, more practice and more strength on wood building design, we will see more and more outstanding wood building designers and excellent wood works.

3) To build a excellent wood building, construction is the key. Normally the construction plays an important role, it will directly affect the appearance, the usability and safety of the building. Compared with the concrete or steel structure, correlation between good project and good construction is higher for wood structure. Elaboration and continuous improvement are the basic guarantees of a excellent wood building.

《木建筑》使用图片授权书

european wood
欧洲木业协会

This letter is to certify that Sino-European Wood Center is entitled to use the selected original articles and pictures from Swedish Wood magazine Trä for the book mù Architecture.

本授权书声明，中欧木结构研究中心有权在《mù Architecture（木建筑）》一书中使用瑞典木业协会杂志《Trä》中的部分原始文章和图片。

Sino-European Wood Center, based on the cooperation between European Wood and Shanghai Jiao Tong University, is an information center to spread knowledge of modern timber structures in China. Swedish Wood, member of European Wood, has been supporting the preparation and publication of mù Architecture by providing articles and pictures of modern timber buildings from Swedish Wood magazine Trä since 2014.

中欧木结构研究中心是基于欧洲木业协会和上海交通大学合作的信息中心，旨在中国传播推广现代木结构的知识。瑞典木业协会作为欧洲木业协会的成员，自2014年以来，一直为《mù Architecture（木建筑）》一书提供现代木结构建筑的相关文章和图片支持，文章和图片来源于瑞典木业协会杂志《Trä》。

All the original articles and pictures from Trä can only be used in China by Sino-European Wood Center for mù Architecture. Science Press is the designated press who is entitled to use the original articles and pictures from Trä only for the publication of mù Architecture.

在中国，所有来源于瑞典木业协会杂志《Trä》的文章和图片，仅中欧木结构研究中心有权用于《mù Architecture（木建筑）》一书。科学出版社作为该书的指定出版社，可以使用来源于瑞典木业协会杂志《Trä》的文章和图片用于《mù Architecture（木建筑）》一书的出版。

All the selected articles and pictures from Trä used in mù

Architecture are listed in the appendix.
所有来源于瑞典木业协会杂志《Trä》的文章和图片详见附件列表。

Date 2017.02.19

Signature JAIN SÖDERLIND

european wood
欧洲木业协会

Appendix: Articles and pictures from Trä used in nn Architecture
附录：所有在《nn Architecture（木建筑）》一书使用的来源于瑞典木业协会杂志《Trä》的文章和图片

Architects promote books	2012.11.22	7
Bishop Edward King Chapel	2013.11.28	10
Multifunctional	2013.12.20	15
Libeskind in Berlin	2013.3.11	10
Showstopping showcases	2013.9.12	8
Tranquil islands in Azerbaijan	2014.11.25	15
Raising the profile of the building's fifth façade	2014.12.1	14
Tribute to mathematics	2014.3.12	10
Diamond acoustics	2014.3.14	10
Right-angled design in recycled wood	2014.3.6	19
Unusual cake shop by Kuma	2014.6.16	11
Preschools embrace wood	2014.6.16	20
Daring French design	2014.9.11	13
Sleep well in the embrace of recycled wood	2014.9.12	14
Individual & industrial	2014.9.17	9
The Nominated for the 2016 Swedish Timber Prize	2015.11.25	10
Privacy and neighbours informed design	2015.12.4	18
Japanese harmony between building and nature	2015.12.7	8
Pine warms futuristic landmark	2015.3.18	17
Expo 2015 – Wood in Milan	2015.9.24	18
Imagination & symmetry come together in material simplicity	2016.3.17	22
Urban development in wood	2016.3.17	13

说明：本辑《木建筑》"简讯"栏目中的8篇文章及"对话"栏目中的访谈1均由瑞典木业协会杂志《Trä》编辑撰写。

Statement: The content in Brief and Interview 1 in Voice are sourced from Swedish Wood magazine Trä in this book.

书籍设计 / M 美光设计

定价:600.00元

(TU-1562.0101)
www.sciencep.com

ISBN 978-7-03-053382-1

9 787030 533821

定价:600.00元

科学出版社互联网入口

赛博古二维码

文物考古分社
部门:(010) 64010983
部门E-mail:arch@mail.sciencep.com